T0205838

SpringerBriefs in Mathematics

SpringerBriefs in Mathematics showcases expositions in all areas of mathematics and applied mathematics. Manuscripts presenting new results or a single new result in a classical field, new field, or an emerging topic, applications, or bridges between new results and already published works, are encouraged. The series is intended for mathematicians and applied mathematicians.

More information about this series at http://www.springer.com/series/10030

Enli Guo • Xiaohuan Mo

The Geometry of Spherically Symmetric Finsler Manifolds

Enli Guo
College of Applied Sciences
Beijing University of Technology
Beijing, Beijing, China

Xiaohuan Mo
School of Mathematical Sciences
Peking University
Beijing, Beijing, China

ISSN 2191-8198 ISSN 2191-8201 (electronic)
SpringerBriefs in Mathematics
ISBN 978-981-13-1597-8 ISBN 978-981-13-1598-5 (eBook)
https://doi.org/10.1007/978-981-13-1598-5

Library of Congress Control Number: 2018948827

Mathematics Subject Classification: 53B40, 53C60, 58E20

This Springer imprint is published by the registered company Springer Nature Singapore Pte Ltd.
The registered company address is: 152 Beach Road, #21-01/04 Gateway East, Singapore 189721, Singapore

Dedicated to the memory of Professor Shiing-Shen Chern

Preface

Spherically symmetric Finsler manifolds are manifolds with spherically symmetric Finsler metrics. Let $\Omega \subseteq \mathbb{R}^n$ is a rotation symmetric domain and F is defined on Ω. F is said to be *spherically symmetric* if the orthogonal group acts as isometries of F. It means that (Ω, F) is invariant under all rotations in $\mathbb{R}^n$. Hence, it is also said to be orthogonally invariant. Such metrics were first studied by Rutz in 1995 who generalizes the classic Birkhoff theorem in general relativity to the Finsler case [63]. Recently, the study of spherically symmetric Finsler metrics has attracted a lot of attention. Many Fisherman geometers have made effort in the study of spherically symmetric Finsler geometry. The classification theorem of projective spherically symmetric metrics of constant flag curvature has been completed [58, 85]. Many new W-quadratic spherically symmetric metrics which are non-trivial are constructed [38, 46].

Spherically symmetric Finsler metrics form a rich class of Finsler metrics. Many classical Finsler metrics with nice curvature properties are spherically symmetric, such as the Bryant metric with one parameter, the metric introduced by Berward, the generalized fourth root metric given by Li-Shen and the Chern-Shen's metric.

Huang-Mo and Zhou independently determined a simple expression of spherically symmetric Finsler metrics (see Sect. 1.3 below). This expression gives us a nice approach to investigate various spherically symmetric metrics, i.e. studying these Finsler metrics with various curvature properties in terms of solving a differential equation or several differential equations. Hence, theory of ODE or PDE is closely related to spherically symmetric Finsler geometry.

The book begins with some basic concepts, examples, propositions, then brings the readers to the most current research areas in geometry of spherically symmetric Finsler manifolds.

In Chap. 1, we give some definitions and examples. We also introduce a basic expression of spherically symmetric metrics.

In Chap. 2, we explicitly construct new examples of spherically symmetric dually flat Finsler metrics by three different approaches.

In Chap. 3, we determine all spherically symmetric Finsler metrics of isotropic Berwald curvature. We also construct explicitly a lot of new isotropic Berwald spherically symmetric Finsler metrics.

In Chap. 4, we obtain the differential equation that characterizes the spherically symmetric Finsler metrics with vanishing Douglas curvature. By solving this equation, we obtain all the spherically symmetric Douglas metric. Many explicit examples are included.

In Chap. 5, we study and characterize (locally) projectively flat spherically symmetric Finsler metrics. We also manufacture new projective spherically symmetric metrics in terms of elementary functions, hypergeometric functions and error functions.

In Chap. 6, we find equations that characterize spherically symmetric metrics of scalar curvature. By using these equations, we construct infinitely many non-projectively flat spherically symmetric of scalar curvature.

In Chap. 7, we study and characterize spherically symmetric Finsler metrics of constant flag curvature. We determine all projectively flat spherically symmetric Finsler metric of negative constant flag curvature. By finding two partial differential equations equivalent to spherically symmetric metrics being of constant flag curvature, we also construct explicitly new spherically symmetric metrics of constant flag curvature.

In the last chapter, we discuss spherically symmetric W-quadratic metrics. In particular, we give a lot of new spherically symmetric Finsler metrics of quadratic Weyl curvature which are non-trivial in the sense that they are not of Weyl type.

Beijing, China
May 2018

Enli Guo
Xiaohuan Mo

Acknowledgements

Authors would like to take this opportunity to thank several people in academic life. First, the first author would like to thank his thesis advisor W. Zhang for the help and advice in Riemannian geometry during his graduate study in Tianjin. The second author would like to thank his thesis advisor Z. Bai and Y. Shen for their help and advice in Riemannian geometry during his graduate study in Hangzhou, and thank another advisor after the thesis, S.S. Chern, for bringing him a wider field – Finsler geometry. The second author thanks Weihuan Chen for providing a good research environment at his current institution. Authors would like to take this opportunity to thank X. Cheng and B. Li for their valuable comments. Authors are also grateful to Y. Li for her generous help.

Contents

Enli Guo is an Associate Professor of College of Mathematics and Physics at the Beijing University of Technology. Prof. Guo received his Ph.D. in mathematics in 2000 from the Chern Institute of Mathematics at Nankai University under the supervision of Professor Weiping Zhang. He got the bachelor degree from the Qufu Normal University and the master degree from the Nankai Institute of Mathematics in China. He works on differential geometry.

Xiaohuan Mo is Professor of Mathematics at Peking University. He obtained his Ph.D. from the Hangzhou University in 1991, with Zhengguo Bai as his advisor. Before coming to Beijing, he did 2 years of postdoctoral studies at the Mathematical Institute of Fudan University in Shanghai. Mo has garnered the award for Natural Science (First Class), the Ministry of Education of P.R. China (2002).

Chapter 1
Spherically Symmetric Finsler Metrics

1.1 Finsler Metrics

By definition, a Finsler metric on a manifold is a family of Minkowski norms on the tangent spaces. A *Minkowski norm* on a vector space V is a nonnegative function $F : V \to [0, +\infty)$ with the following properties:

(i) F is positively y-homogeneous of degree one, i.e. for any $y \in V$ and any $\lambda > 0$,

$$F(\lambda y) = \lambda F(y);$$

(ii) F is C^∞ on $V \backslash \{0\}$ and for any tangent vector $y \in V \backslash \{0\}$, the following bilinear symmetric form $g_y(u, v) : V \times V \longrightarrow \mathbb{R}$ is positive definite:

$$g_y(u, v) := \frac{1}{2} \frac{\partial^2}{\partial s \partial t} [F^2(y + su + tv)] \mid_{s=t=0} .$$

Let $\langle \, , \, \rangle$ denote the standard inner product on $\mathbb{R}^n$, defined by $\langle u, v \rangle := \sum_{i=1}^n u^i v^i$. Then $|y| := \sqrt{\langle y, y \rangle}$ is called the *standard Euclidean norm* on $\mathbb{R}^n$.

Let M be a differentiable manifold. Let $TM = \bigcup_{x \in M} T_x M$ be the tangent bundle on M, where $T_x M$ is tangent space at $x \in M$. We denote a typical point in TM by (x, y), where $y \in T_x M$, and set $TM_0 := TM \backslash \{0\}$ where $\{0\}$ stands for $\{(x, 0) \mid x \in M, 0 \in T_x M\}$. A *Finsler metric* on M is a function $F : TM \to [0, \infty)$ with the following properties:

(a) F is C^∞ on TM_0;
(b) At each point $x \in M$, the restriction $F_x := F|_{T_x M}$ is a Minkowski norm on $T_x M$.

The pair (M, F) is called a *Riemann-Finsler manifold* or *Finsler manifold* for short.

E. Guo, X. Mo, *The Geometry of Spherically Symmetric Finsler Manifolds*, SpringerBriefs in Mathematics, https://doi.org/10.1007/978-981-13-1598-5_1

Let (M, F) be a Finsler manifold. Let (x^i, y^i) be a standard local coordinate system in TM, i.e., y^i's are determined by $y = y^j(\partial/\partial x^j)|_x$. For a vector $y = y^j(\partial/\partial x^j)|_x \neq 0$, let $g_{ij}(x, y) := \frac{1}{2}(F^2)_{y^i y^j}(x, y)$. The *fundamental tensor* $\mathcal{G}$ is defined by

$$\mathcal{G} := g_{ij} dx^i \otimes dx^j.$$

By the homogeneity of F,

$$F(x, y) = \sqrt{g_{ij}(x, y) y^i y^j}.$$

A Finsler metric $F = F(x, y)$ is called a *Riemannian metric* if the $g_{ij} = g_{ij}(x)$ are functions of $x \in M$ only.

There are three special Riemannian metrics.

Example 1.1.1 (Euclidean metric) The simplest metric is the Euclidean metric $\alpha_0 = \alpha_0(x, y)$ on $\mathbb{R}^n$, which is defined by

$$\alpha_0(x, y) := |y|, \quad y = (y^i) \in T_x\mathbb{R}^n \cong \mathbb{R}^n.$$

We will simply denote $(\mathbb{R}^n, \alpha_0)$ by $\mathbb{R}^n$, which is called *Euclidean space*.

Example 1.1.2 (Spherical metric) Let $S^n := \{x \in \mathbb{R}^{n+1} | |x| = 1\}$ denote the standard unit sphere in $\mathbb{R}^{n+1}$ in a natural way. The induced metric α_+ on S^n is defined by $\alpha_+ = \|y\|_x$, for $y \in T_x S^n \subset \mathbb{R}^{n+1}$, where $\| \cdot \|_x$ denotes the induced Euclidean norm on $T_x S^n$. Let $\varphi : \mathbb{R}^n \to S^n \subset \mathbb{R}^{n+1}$ be defined by

$$\varphi(x) := \left(\frac{x}{\sqrt{1 + |x|^2}}, \frac{1}{\sqrt{1 + x^2}} \right).$$

Then φ pulls back α_+ on the upper hemisphere to a Riemannian metric on $\mathbb{R}^n$, which is given by

$$\alpha_+ = \frac{\sqrt{|y|^2 + (|x|^2|y|^2 - \langle x, y \rangle^2)}}{1 + |x|^2}, \quad y \in T_x\mathbb{R}^n \cong \mathbb{R}^n.$$

Example 1.1.3 (Hyperbolic metric) Let $\mathbb{B}^n$ denote the unit ball in $\mathbb{R}^n$. Define

$$\alpha_{-1} := \frac{\sqrt{|y|^2 - (|x|^2|y|^2 - \langle x, y \rangle^2)}}{1 - |x|^2}, \quad y \in T_x\mathbb{B}^n \cong \mathbb{R}^n.$$

We call α_{-1} the *Klein metric* and denote $(\mathbb{B}^n, \alpha_{-1})$ by $\mathbb{H}^n$.

The Riemannian metrics in Examples 1.1.1, 1.1.2 and 1.1.3 can be expressed in one single formula

$$\alpha_\mu := \frac{\sqrt{|y|^2 + \mu(|x|^2|y|^2 - \langle x, y\rangle^2)}}{1 + \mu|x|^2}, \quad y \in T_x\mathbb{B}^n(r_\mu) \cong \mathbb{R}^n$$

where $r_\mu := 1/\sqrt{-\mu}$ if $\mu < 0$ and $r_\mu := +\infty$ if $\mu \geq 0$. The metric α_μ can be expressed as $\alpha_\mu = \sqrt{a_{ij}y^iy^j}$, where

$$a_{ij} = \frac{1}{1 + \mu|x|^2}\left(\delta_{ij} - \frac{\mu x^i x^j}{1 + \mu|x|^2}\right).$$

A Finsler metric $F = F(x, y)$ is called a *locally Minkowski metric* if the $g_{ij} = g_{ij}(y)$ are functions of y only.

Example 1.1.4 Given a Minkowski norm $\phi : V \to [0, +\infty)$ on a vector space V, one can construct $\Omega := \{v \in V|\phi(v) < 1\}$. A domain Ω in V defined by a Minkowski norm ϕ is called a *strongly convex domain*. Thus $F(x, y)$ is a (locally) Minkowski metric on Ω, where $F(x, y) := \phi(y)$.

Let $\alpha = \sqrt{a_{ij}(x)y^iy^j}$ be a Riemannian metric on a manifold M and $\beta = b_i(x)y^i$ be a 1-form on M. Let

$$\|\beta_x\|_\alpha := \sup_{y \in T_xM} \frac{\beta(x, y)}{\alpha(x, y)} = \sqrt{a^{ij}(x)b_i(x)b_j(x)}.$$

It is easy to show that $F := \alpha + \beta$ is a Finsler metric if and only if $\|\beta_x\|_\alpha < 1$ for any $x \in M$. The Finsler metric $F = \alpha + \beta$ with $sup_{x\in M}\|\beta_x\|_\alpha < 1$ is called a *Randers metric* on M.

Example 1.1.5 (Funk metric) Let

$$F := \frac{\sqrt{|y|^2 - (|x|^2|y|^2 - \langle x, y\rangle^2)} + \langle x, y\rangle}{1 - |x|^2}, \quad y \in T_x\mathbb{B}^n \cong \mathbb{R}^n.$$

$F = F(x, y)$ is a Randers metric on $\mathbb{B}^n$, called the *Funk metric* on $\mathbb{B}^n$.

For an arbitrary constant vector $a \in \mathbb{R}^n$ with $|a| < 1$, let

$$F_a := \frac{\sqrt{|y|^2 - (|x|^2|y|^2 - \langle x, y\rangle^2)} + \langle x, y\rangle}{1 - |x|^2} + \frac{\langle a, y\rangle}{1 + \langle a, x\rangle},$$

where $y \in T_x\mathbb{B}^n \cong \mathbb{R}^n$. $F_a = F_a(x, y)$ is a Randers metric on $\mathbb{B}^n$. We call F_a the *generalized Funk metric* on $\mathbb{B}^n$. Note that $F_0 = F$ is the Funk metric on $\mathbb{B}^n$.

Randers metric were first studied by physicist G. Randers, in 1941 from the standard point of general relativity [62]. Later on, these metrics were applied to the theory of the electron microscope by R. S. Ingarden in 1957, who first named them Randers metrics.

1.2 Spherically Symmetric Finsler Metrics

Spherically symmetric metrics are Finsler metrics with orthogonal invariance. Such metrics were first studied by S. F. Rutz in [43, 63]. The Euclidean metric, the spherical metric, the Klein metric and the Funk metric are special spherically symmetric metrics.

Definition 1.2.1 Let F be a Finsler metric on $\mathbb{B}^n(r_\mu)$. F is said to be *spherically symmetric* (*orthogonally invariant* in an alternative terminology in [25, 30]) if it satisfies

$$F(Ax, Ay) = F(x, y) \tag{1.1}$$

for all $x \in \mathbb{B}^n(r_\mu)$, $y \in T_x\mathbb{B}^n(r_\mu)$ and $A \in O(n)$.

Note that a Finsler metric F is spherically symmetric if and only if the orthogonal group $O(n)$ acts as isometries of F. In general, the domain of a spherically symmetric metric can be an annuli, a ball or the entire space $\mathbb{R}^n$.

Example 1.2.1 ([6]) Let

$$F := \frac{\left(\sqrt{|y|^2 - (|x|^2|y|^2 - \langle x, y\rangle^2)} + \langle x, y\rangle\right)^2}{(1 - |x|^2)^2\sqrt{|y|^2 - (|x|^2|y|^2 - \langle x, y\rangle^2)}}, \tag{1.2}$$

where $y \in T_x\mathbb{B}^n \cong \mathbb{R}^n$. Then $F = F(x, y)$ is a spherically symmetric Finsler metric on $\mathbb{B}^n$.

Let $\bar{F}$ be the Funk metric on $\mathbb{B}^n$ defined in Example 1.1.5. What is interesting to us is the following relationship between $\bar{F}$ and F. Let

$$\bar{\alpha} := \frac{\sqrt{|y|^2 - (|x|^2|y|^2 - \langle x, y\rangle^2)}}{1 - |x|^2}, \quad \bar{\beta} := \frac{\langle x, y\rangle}{1 - |x|^2}$$

and $\lambda := \frac{1}{1-|x|^2}$. Then $\bar{F} = \bar{\alpha} + \bar{\beta}$, and

$$F = \frac{(\alpha + \beta)^2}{\alpha} = \alpha + 2\beta + \frac{\beta^2}{\alpha}$$

where $\alpha := \lambda\bar{\alpha}$ and $\beta = \lambda\bar{\beta} = \frac{1}{2}d\lambda$. Finsler metrics in the form $F = \frac{(\alpha+\beta)^2}{\alpha}$ are called *square metrics* [75, 76, 84].

Example 1.2.2 In [61], A. V. Pogorelov constructed the following spherically symmetric Finsler metric:

$$F = \frac{1}{3|y|}[(3 + |x|^2)|y|^2 + \langle x, y\rangle^2].$$

where $x = (x^1, x^2)$ and $y = (y^1, y^2)$ (see Exercise 9.3. in [41]).

Example 1.2.3 ([70]) Let ε be an arbitrary number with $\varepsilon < 1$. Let

$$F_\varepsilon = \frac{\sqrt{|y|^2 - (|x|^2|y|^2 - \langle x, y\rangle^2)} + \langle x, y\rangle}{2(1 - |x|^2)} - \varepsilon \frac{\sqrt{|y|^2 - \varepsilon^2(|x|^2|y|^2 - \langle x, y\rangle^2)} + \varepsilon\langle x, y\rangle}{2(1 - \varepsilon^2|x|^2)}$$

where $y \in T_x\mathbb{B}^n \cong \mathbb{R}^n$. F_ε is a spherically symmetric metric on $\mathbb{B}^n$.

Note that F_ε is no longer of Randers type if $\varepsilon \neq 0, -1$ and $F_{-1} = \alpha_{-1}$ is the Klein metric on $\mathbb{B}^n$. Let $\bar{F}$ be the Funk metric on $\mathbb{B}^n$ defined in Example 1.1.5. We can express F_ε by

$$F_\varepsilon = \frac{1}{2}\left[\bar{F}(x, y) - \varepsilon\bar{F}(\varepsilon x, y)\right].$$

Example 1.2.4 Let $F = \sqrt{\sqrt{A} + B}$ be a generalized fourth root metric on $\mathbb{B}^n \subset \mathbb{R}^n$ defined by [37].

$$A := \frac{|y|^4 + (|x|^2|y|^2 - \langle x, y\rangle^2)^2}{4(1 + |x|^4)^2},$$

$$B := \frac{(1 + |x|^4)|x|^2|y|^2 + (1 - |x|^4)\langle x, y\rangle^2}{2(1 + |x|^4)^2}.$$

Then F is a spherically symmetric metric.

Example 1.2.5 ([8]) Let ε be an arbitrary number with $|\varepsilon| < 1$. Let

$$F_\varepsilon := \frac{1}{\Psi}\left\{\sqrt{\Psi[\frac{1}{2}(\sqrt{\Phi^2 + (1 - \varepsilon^2)|y|^4} + \Phi)]} + \sqrt{1 - \varepsilon^2}\langle x, y\rangle\right\}$$

where

$$\Phi := \varepsilon|y|^2 + |x|^2|y|^2 - \langle x, y\rangle^2, \quad \Psi := 1 + 2\varepsilon|x|^2 + |x|^4.$$

$F_\varepsilon = F_\varepsilon(x, y)$ is a spherically symmetric metric on $\mathbb{R}^n$. Note that if $\varepsilon = 1$, then $F_1 = \alpha_{+1}$ is the spherical metric on $\mathbb{R}^n$.

The spherically symmetric metrics F_ε in Example 1.2.5 is a special family of Bryant's metrics expressed in a local coordinate system [7].

We will discuss geometric properties of Examples 1.2.1, 1.2.2, 1.2.3, 1.2.4 and 1.2.5 later.

1.3 An Expression of Spherically Symmetric Metrics

In this section, we are going to determine an expression of spherically symmetric metrics. We will see that all spherically symmetric metrics are the so-called general (α, β)-metrics (for exact definition, see page 8 or [82]).

Let $|\cdot|$ and $\langle \ , \ \rangle$ be the standard Euclidean norm and inner product on $\mathbb{R}^n$.

Proposition 1.3.1 *A Finsler metric F on $\mathbb{B}^n(r_\mu)$ is spherically symmetric if and only if there is a function $\phi : [0, r_\mu) \times \mathbb{R} \to \mathbb{R}$ such that*

$$F(x, y) = |y|\phi\left(|x|, \frac{\langle x, y\rangle}{|y|}\right) \tag{1.3}$$

where $(x, y) \in T\mathbb{B}^n(r_\mu) \setminus \{0\}$. In particular, all spherically symmetric Finsler metrics are general (α, β)-metrics.

Proof Assume that $F(x, y) = |y|\phi(|x|, \langle x, y\rangle/|y|)$ for some $\phi : [0, r_\mu) \times \mathbb{R} \to \mathbb{R}$. It is easy to see

$$\langle Ax, \ Ay\rangle = \langle x, \ A^T Ay\rangle = \langle x, \ y\rangle$$

for $x, y \in \mathbb{R}^n$ and $A \in O(n)$. In particular, $|Ax| = |x|$ for $x \in \mathbb{R}^n$. Hence

$$F(Ax, \ Ay) = |Ay|\phi\left(|Ax|, \frac{\langle Ax, Ay\rangle}{|Ay|}\right) = |y|\phi\left(|x|, \frac{\langle x, y\rangle}{|y|}\right) = F(x, y).$$

Conversely, suppose that F is orthogonally invariant. Denote by $e_1, \ldots, e_n$ the standard orthonormal basis of $\mathbb{R}^n$, where

$$e_j = (0, \ldots, 0, \underset{j}{1}, 0, \ldots, 0), \quad j = 1, \ldots, n. \tag{1.4}$$

Put

$$\epsilon_1 = \frac{x}{|x|}, \quad \epsilon_2 = \frac{y - \frac{\langle y,x\rangle}{|x|^2}x}{\left|y - \frac{\langle y,x\rangle}{|x|^2}x\right|}. \tag{1.5}$$

Then ϵ_1 and ϵ_2 are orthonormal vectors in $\mathbb{R}^n$. It follows that there exists an $A \in O(n)$ such that

$$A\epsilon_1 = e_1, \quad A\epsilon_2 = e_2. \tag{1.6}$$

A simple calculation gives

$$\left|y - \frac{\langle y, x\rangle}{|x|^2}x\right|^2 = |y|^2 - \frac{\langle x, y\rangle^2}{|x|^2}. \tag{1.7}$$

By using the first formula of (1.5) and the first formula of (1.6) we obtain

$$Ax = A(|x|\epsilon_1) = |x|A\epsilon_1 = |x|e_1. \tag{1.8}$$

Together with (1.7), the second formula of (1.5) and the second formula of (1.6) we get

$$\begin{aligned}
Ay &= A\left(|y - \frac{\langle y, x\rangle}{|x|^2}x|\epsilon_2 + \frac{\langle y, x\rangle}{|x|^2}x\right) \\
&= A\left(\frac{\langle x, y\rangle}{|x|^2}x + \frac{\sqrt{|x|^2|y|^2 - \langle x, y\rangle^2}}{|x|}\epsilon_2\right) \\
&= \frac{\langle x, y\rangle}{|x|^2}Ax + \frac{\sqrt{|x|^2|y|^2 - \langle x, y\rangle^2}}{|x|}A\epsilon_2 \\
&= \frac{\langle x, y\rangle}{|x|}e_1 + \frac{\sqrt{|x|^2|y|^2 - \langle x, y\rangle^2}}{|x|}e_2.
\end{aligned} \tag{1.9}$$

Applying the orthogonal invariance of F we obtain

$$\begin{aligned}
F(x, y) &= F(Ax,\ Ay) \\
&= F\left(|x|e_1, \frac{\langle x, y\rangle}{|x|}e_1 + \frac{\sqrt{|x|^2|y|^2 - \langle x, y\rangle^2}}{|x|}e_2\right) \\
&= F\left(|x|, 0, \ldots, 0; \frac{\langle x, y\rangle}{|x|}, \frac{\sqrt{|x|^2|y|^2 - \langle x, y\rangle^2}}{|x|}, 0, \ldots 0\right) \\
&= \psi(|x|, \langle x, y\rangle, |y|)
\end{aligned} \tag{1.10}$$

where $\psi : [0, r_\mu) \times \mathbb{R}^2 \to \mathbb{R}$ and we have used (1.4), (1.8), (1.9). Note that F is homogeneous of degree one with respect to y. Hence

$$\begin{aligned}
\lambda\psi(|x|, \langle x, y\rangle, |y|) &= \lambda F(x, y) = F(x, \lambda y) \\
&= \psi(|x|, \langle x, \lambda y\rangle, |\lambda y|) = \psi(|x|, \lambda\langle x, y\rangle, \lambda|y|)
\end{aligned}$$

for $\lambda \in [0, \infty)$. In particular,

$$\frac{1}{|y|}\psi(|x|, \langle x, y\rangle, |y|) = \psi\left(|x|, \frac{\langle x, y\rangle}{|y|}, 1\right) := \phi\left(|x|, \frac{\langle x, y\rangle}{|y|}\right)$$

where $y \in T_x\mathbb{B}^n(r_\mu) \setminus \{0\}$ and $\phi : [0, r_\mu) \times \mathbb{R} \to \mathbb{R}$. Plugging this into (1.10) yields (1.3). □

A Finsler metric on a manifold M in the following form is said to be *general* (α, β) *type*

$$F = \alpha\phi\left(r^2, \frac{\beta}{\alpha}\right)$$

where α is a Riemannian metric, β is a 1-form on M, $r = \|\beta\|_\alpha$ and $\phi(r^2, s)$ is a C^∞ function satisfying (see [82])

$$\phi(s) - s\phi_s(s) > 0, \quad \phi(s) - s\phi_s(s) + (r^2 - s^2)\phi_{ss}(s) > 0, \quad |s| \leq r < b_o$$

where $n \geq 3$ or

$$\phi(s) - s\phi_s(s) + (r^2 - s^2)\phi_{ss}(s) > 0, \quad |s| \leq r < b_o$$

where $n = 2$.

Example 1.3.1 Let F be the Funk metric on $\mathbb{B}^n$ defined in Example 1.1.5. Then F can also be expressed in the form

$$F = |y|\phi(r, s), \quad r := |x|, \quad s := \frac{\langle x, y\rangle}{|y|}$$

and

$$\phi(r, s) = \frac{\sqrt{1 - r^2 + s^2}}{1 - r^2} + \frac{s}{1 - r^2}. \tag{1.11}$$

Example 1.3.2 It is clear that the corresponding function $\phi(r, s)$ of (1.2) is given by

$$\phi = \frac{(\sqrt{1 - r^2 + s^2} + s)^2}{(1 - r^2)^2\sqrt{1 - r^2 + s^2}} \tag{1.12}$$

We notice that the above ϕ in (1.11) and (1.12) satisfies the following PDE

$$r\phi_{ss} + s\phi_{rs} - \phi_r = 0.$$

Here ϕ_r means the derivation of ϕ with respect to the first variable r. This is indeed an amazing phenomenon.

Formula (1.3) gives us a nice approach to investigate all spherically symmetric metrics. In order to study spherically symmetric metrics we only focus on the corresponding function ϕ.

Chapter 2
Dually Flat Spherically Symmetric Metrics

Dually flat Finsler metrics arise from α-flat information structures on Riemann-Finsler manifolds. Such Finsler metrics was introduced by Amari-Nagaoka and Z. Shen (See [3, 72]). Recently the study of dually flat Finsler metrics has attracted a lot of attention [14, 15, 29, 30, 33, 39, 79, 81].

2.1 Definition and Some Explicit Constructions

A Finsler metric $F = F(x, y)$ on an open subset $\mathscr{U} \subset \mathbb{R}^m$ is *dually flat* if F satisfies

$$(F^2)_{x^i y^j} y^i = 2(F^2)_{x^j} \tag{2.1}$$

where $x = (x^1, \cdots, x^m) \in \mathscr{U}$, and $y = y^j \frac{\partial}{\partial x^j}|_x \in T_x\mathscr{U}$.

Example 2.1.1 Consider a Minkowski norm $\varphi : \mathbb{R}^m \to \mathbb{R}$ on $\mathbb{R}^m$. One can construct the strongly convex domain Ω and the Minkowski metric Φ as following:

$$\Omega := \{v \in \mathbb{R}^m | \varphi(v) < 1\}, \quad \Phi(x, y) = \varphi(y)$$

(see Example 1.1.4). By using Φ and the homothetic field $\mathscr{V}$, we produce the *Funk metric F on the strongly convex domain Ω* in terms of navigation problem where $\mathscr{V}_x := x$ is a radical vector field on Ω satisfying $\Phi(x, \mathscr{V}_x) = \varphi(x) < 1$ [26]. F is dually flat [14]. In particular, for $\varphi(y) = |y|$, we get the Funk metric Θ on the unit ball $\mathbb{B}^m \subset \mathbb{R}^m$:

$$\Theta = \frac{\sqrt{|y|^2 - (|x|^2|y|^2 - \langle x, y\rangle^2)}}{1 - |x|^2} + \frac{\langle x, y\rangle}{1 - |x|^2}$$

E. Guo, X. Mo, *The Geometry of Spherically Symmetric Finsler Manifolds*, SpringerBriefs in Mathematics, https://doi.org/10.1007/978-981-13-1598-5_2

where $y \in T_x\mathbb{B}^m \cong \mathbb{R}^m$. The Funk metric Θ can be expressed in the form $\Theta = \sqrt{\Theta_1^2 + \Theta_2^2}$ where

$$\Theta_1 = |y|\sqrt{g(t) + g'(t)s^2}, \quad \Theta_2 = |y|[h(t)s^2 + \frac{1}{6}h'(t)s^4]^{\frac{1}{4}}$$

where

$$g(t) = \frac{1}{1-2t}, \quad h(t) = g(t)^2, \quad t = \frac{|x|^2}{2}, \quad s = \frac{\langle x, y\rangle}{|y|}.$$

We can verify that Θ_1 and Θ_2 satisfy (2.1) by straightforward calculations.

Conversely, it is easy to see that if F_1 and F_2 satisfy (2.1) then $\sqrt{aF_1^2 + bF_2^2}$ is also a solution of (2.1) where a and b are non-negative constants.

After noting this interesting fact, we manufacture explicitly new dually flat Finsler metrics in this section. These Finsler metrics contains the Funk metric Θ on the unit ball $\mathbb{B}^m$.

Our approach is to discuss the solution of dually flat equation (2.1) in the following forms

$$F(x, y) = |y|\sqrt{\sum_{j=0}^{l} f_j\left(\frac{|x|^2}{2}\right)\frac{\langle x, y\rangle^j}{|y|^j}} \tag{2.2}$$

and

$$F(x, y) = |y|\left[\sum_{j=0}^{l} f_j\left(\frac{|x|^2}{2}\right)\frac{\langle x, y\rangle^j}{|y|^j}\right]^{\frac{1}{4}}. \tag{2.3}$$

In fact, (2.2) (resp. (2.3)) contains Θ_1 (resp. Θ_2).

First we determine general solutions of (2.1) in the form (2.2) (Proposition 2.1.2). Secondly, we discuss necessary and sufficient condition on f_j for (2.3) to satisfy (2.1) (see Proposition 2.1.3 below). In particular, we obtain some solutions of (2.1) when $l = 4$ and $f_1 = f_3 = 0$ (Proposition 2.1.4). Finally, using the fundamental property of (2.1) we construct a lot of new dually flat Finsler metrics. Precisely, we show the following:

Theorem 2.1.1 *Let $f(t, s)$ be a function defined by*

$$\begin{aligned} f(t, s) &= g(t) + h(t)s + g'(t)s^2 + \tfrac{1}{6}h'(t)s^3 \\ &\quad + \sum_{j=2}^{n}(-1)^{j-1}\tfrac{(2j-3)!!}{(2j+1)!}h^{(j)}(t)s^{2j+1} + b\sqrt{\tfrac{-s^2}{(c+t)^3} + \tfrac{s^4}{2(c+t)^4}} \end{aligned} \tag{2.4}$$

where b, c are constants and g is an any differentiable function and h is an any polynomial function of N degree where $N \leq n$ and $h^{(j)}$ denotes j-order derivative for $h(t)$. Then the following spherically symmetric Finsler metric on $\mathbb{B}^m(r_\mu)$

$$F = |y|\sqrt{f\left(\frac{|x|^2}{2}, \frac{\langle x, y\rangle}{|y|}\right)} \tag{2.5}$$

is dually flat.

When $b = \frac{1}{\sqrt{2}}$, $c = -\frac{1}{2}$, $h(t) = 0$ and $g(t) = \frac{1}{1-2t}$, (2.5) is reduced to the famous Funk metric Θ on the unit ball $\mathbb{B}^m$.

By using Propositions 2.1.2 and 2.1.4 below, we obtain more dually flat spherically symmetric metrics (Theorem 2.1.2).

2.1.1 Some Lemmas

Let $F = |y|\phi\left(\frac{|x|^2}{2}, \frac{\langle x, y\rangle}{|y|}\right)$ be a spherically symmetric Finsler metric on $\mathbb{B}^m(r_\mu)$. Let

$$r := |y|, \quad t := \frac{|x|^2}{2}, \quad s := \frac{\langle x, y\rangle}{|y|}, \tag{2.6}$$

$$r^i := r_i := \frac{y^i}{|y|}, \quad x_i := x^i, \quad s^i := s_i := x_i - sr_i. \tag{2.7}$$

Direct computations yield that

$$t_{x^i} = x^i = s_i + sr_i, \quad s_{x^i} = r_i, \quad r_{y^i} = r_i, \quad s_{y^i} = \frac{s_i}{r} \tag{2.8}$$

where we have used (2.6) and (2.7).

Lemma 2.1.1 *Let $f = f(r, t, s)$ be a function on a domain $\mathscr{U} \subset \mathbb{R}^3$. Then*

$$f_{x^i} = (r_i, s_i)\begin{pmatrix} f_s + sf_t \\ f_t \end{pmatrix}, \quad f_{y^i} = (r_i, s_i)\begin{pmatrix} f_r \\ f_s/r \end{pmatrix}. \tag{2.9}$$

Proof By (2.6), (2.7) and (2.8), we have (2.9).

Corollary 2.1.1 *Let $F = |y|\phi\left(\frac{|x|^2}{2}, \frac{\langle x, y\rangle}{|y|}\right)$ be a spherically symmetric Finsler metric on $\mathbb{B}^m(r_\mu)$. Then*

$$F_{x^i} = r\left[\phi_t s_i + (\phi_s + s\phi_t)r_i\right], \tag{2.10}$$

$$F_{y^i} = \phi r_i + \phi_s s_i. \tag{2.11}$$

In particular

$$F_0 = r^2 \cdot (\phi_s + s\phi_t) \tag{2.12}$$

where $F_0 =: F_{x^i} y^i$.

Proof In fact, we have the following

$$F_r = \phi, \quad F_t = r\phi_t, \quad F_s = s\phi_s.$$

It follows from (2.9) that (2.10) and (2.11) hold. Note that s_i and r_i are positively homogeneous of degree 0 and 1 respectively. Hence $s_i y^i = 0,\ r_i y^i = r$. Combine these with (2.10) we have (2.12).

Remark 2.1.1 By using (2.10), (2.12) and the following modified Hamel equation,

$$(F_0)_{y^i} = 2F_{x^i},$$

we see that F is projectively flat (for definition, see (2.71) below) if and only if ϕ satisfies $s\phi_{ts} + \phi_{ss} - \phi_t = 0$.

Putting

$$\bar{h}^i_j = \delta^i_j - r^i r_j \tag{2.13}$$

where r^i and r_j are defined in (2.7). From (2.7) and (2.13) we have

$$[r^i]_{x^j} = 0, \quad [r^i]_{y^j} = \frac{1}{r}\bar{h}^i_j. \tag{2.14}$$

Together with (2.6), (2.7) and (2.8), we have

$$[s^i]_{x^j} = \bar{h}^i_j, \quad [s^i]_{y^j} = -\frac{1}{r}(s_j r^i + s\bar{h}^i_j). \tag{2.15}$$

By direct calculations one obtains $\phi_{y^j} = \frac{1}{r}\phi_s s_j$, $(\phi_s)_{y^j} = \frac{1}{r}\phi_{ss} s_j$. Together with (2.11), (2.14) and (2.15) we get

$$F_{y^i y^j} = \frac{1}{r}\left[\phi_s r_i s_j + \phi\bar{h}^i_j + \phi_{ss} s_i s_j - \phi_s(s_j r_i + s\bar{h}^i_j)\right]. \tag{2.16}$$

Lemma 2.1.2 *Let* $F = |y|\phi\left(\frac{|x|^2}{2}, \frac{\langle x, y\rangle}{|y|}\right)$ *be a spherically symmetric Finsler metric on* $\mathbb{B}^m(r_\mu)$. *Then the coefficients of the fundamental tensor* g_{ij} *are given by*

$$g_{ij} = \phi c_0 \delta_{ij} + (r_i, s_i) X_1 \begin{pmatrix} r_j \\ s_j \end{pmatrix} \tag{2.17}$$

where $c_0 = \phi - s\phi_s$ *and*

$$X_1 = \begin{pmatrix} s\phi\phi_s & \phi\phi_s \\ \phi\phi_s & \phi_s^2 + \phi\phi_{ss} \end{pmatrix}. \tag{2.18}$$

Proof (2.17) is an immediate conclusion of (2.11) and (2.16).

Setting $S := \Sigma_j \begin{pmatrix} r_j \\ s_j \end{pmatrix} (r_j, s_j)$. Then

$$S = \Sigma_j \begin{pmatrix} r_j r^j & r_j s^j \\ s_j r^j & s_j s^j \end{pmatrix} = \begin{pmatrix} 1 & 0 \\ 0 & 2t - s^2 \end{pmatrix}. \tag{2.19}$$

Note that all spherically symmetric Finsler metrics are general (α, β)-metrics. By using Proposition 3.3 in [82] (also see Sect. 1.3), we have $c_0 > 0$ and

$$\Delta := \phi - s\phi_s + (2t - s^2)\phi_{ss} > 0. \tag{2.20}$$

Using (2.17) and (2.19), we obtain that the inverse matrix $(g^{ij}) = (g_{ij})^{-1}$ is given by

$$g^{ij} = \frac{1}{c_0 \phi} \delta^{ij} - (r^i, s^i) X_2 \begin{pmatrix} r^j \\ s^j \end{pmatrix} \tag{2.21}$$

where

$$X_2 = \frac{1}{c_0 \phi^3 \Delta} \begin{pmatrix} \phi_s [s\phi\Delta - (2t - s^2)\phi_s c_0] & \phi\phi_s c_0 \\ \phi\phi_s c_0 & \phi^2 \phi_{ss} \end{pmatrix}.$$

Let

$$G_j = \frac{1}{4} \left[y^k (F^2)_{y^j x^k} - (F^2)_{x^j} \right].$$

A direct calculation yields

$$G_j = \frac{1}{2} \left[(F F_0)_{y^j} - (F^2)_{x^j} \right]. \tag{2.22}$$

Using (2.12), we obtain

$$(F F_0)_r = 3r^2 \phi(\phi_s + s\phi_t), \quad (F F_0)_s = r^3 \left[\phi_s(\phi_s + s\phi_t) + \phi(\phi_{ss} + \phi_t + s\phi_{ts}) \right].$$

Together with Lemma 2.1.2, we have

$$(FF_0)_{y^i} = (r_i, s_i) \begin{pmatrix} 3r^2\phi(\phi_s + s\phi_t) \\ r^2[\phi_s(\phi_s + s\phi_t) + \phi(\phi_{ss} + \phi_t + s\phi_{ts})] \end{pmatrix}. \tag{2.23}$$

Similarly, we get

$$(F^2)_{x^i} = (r_i, s_i) \begin{pmatrix} 2r^2\phi\phi_s + 2sr^2\phi\phi_t \\ 2r^2\phi\phi_t \end{pmatrix}. \tag{2.24}$$

Substituting (2.23) and (2.24) into (2.22) we have

$$G_i = \frac{r^2}{2}(r_i, s_i) \begin{pmatrix} c_1 \\ c_2 \end{pmatrix} \tag{2.25}$$

where

$$c_1 := \phi(\phi_s + s\phi_t), \tag{2.26}$$

$$c_2 := s(\phi_s\phi_t + \phi\phi_{ts}) + \phi_s^2 + \phi\phi_{ss} - \phi\phi_t. \tag{2.27}$$

We will use (2.25) and (2.27) in Sect. 2.1.3.

The following lemma will be used in the proof of Proposition 2.1.3. The proof is omitted.

Lemma 2.1.3 *We have*

$$\text{(i)} \ \sum_{i=1}^m a_i \sum_{j=1}^m b_j = \sum_{k=1}^{2m} \sum_{i+j=k} a_i b_j,$$

$$\text{(ii)} \ \sum_{i=1}^m a_i \sum_{j=1}^m b_j - \sum_{i=1}^m c_i \sum_{j=1}^m d_j = \sum_{k=1}^{2m} \sum_{i+j=k} (a_i b_j - c_i d_j). \tag{2.28}$$

2.1.2 Dually Flat Finsler Metrics

In this subsection, we are going to establish the partial differential equation for a spherically symmetric Finsler metric F to be dually flat.

Recall that a Finsler metric $F = F(x, y)$ on an open subset $\mathcal{U} \subset \mathbb{R}^m$ is *dually flat* if and only if it satisfies (2.1) [14]. Note that $(F^2)_{x^i y^j} = (F^2)_{y^j x^i}$. It follows that

$$G_j = \frac{1}{4}[(F^2)_{x^i y^j} y^i - (F^2)_{x^j}]. \tag{2.29}$$

By (2.1) and (2.29), $F = F(x, y)$ is dually flat if and only if

$$4G_j = (F^2)_{x^j}. \tag{2.30}$$

Consider spherically symmetric Finsler metric $F(x, y) = r\phi(t, s)$ where r, t and s satisfy (2.6). By (2.24) and (2.25), (2.30) holds if and only if

$$\begin{aligned}(r_j, s_j)\begin{pmatrix} c_1 \\ c_2 \end{pmatrix} &= \phi[\phi_t s_j + (\phi_s + s\phi_t) r_j] \\ &= (r_j, s_j)\begin{pmatrix} \phi(\phi_s + s\phi_t) \\ \phi\phi_t \end{pmatrix}.\end{aligned} \tag{2.31}$$

Taking x and y with $x \wedge y \neq 0$, we obtain

$$\begin{aligned}\sum(r_j)^2 \sum(s_j)^2 - (\sum r_j s_j)^2 &= |y|^2 \left[\sum(x^j - \tfrac{\langle x,y\rangle}{|y|^2} y^j)\right] - \sum \tfrac{y^j}{|y|}\left(x^j - \tfrac{\langle x,y\rangle}{|y|^2} y^j\right) \\ &= |y|^2|x|^2 - \langle x, y\rangle^2 > 0.\end{aligned}$$

It follows that $(r_1, \cdots, r_m)$ and $(s_1, \cdots, s_m)$ are not collinear. Together with (2.31) we get

$$c_2 = \phi\phi_t. \tag{2.32}$$

Combining this with (2.27), we get

$$s(\phi_s\phi_t + \phi\phi_{ts}) + \phi_s^2 + \phi\phi_{ss} - 2\phi\phi_t = 0. \tag{2.33}$$

Define $f = \phi^2$. Then

$$f_t = 2\phi\phi_t, \tag{2.34}$$

$$f_{ts} = 2(\phi_s\phi_t + \phi\phi_{ts}), \tag{2.35}$$

$$f_{ss} = 2(\phi_s^2 + \phi\phi_{ss}). \tag{2.36}$$

Plugging (2.34), (2.35) and (2.36) into (2.33) yields

$$sf_{ts} + f_{ss} - 2f_t = 0. \tag{2.37}$$

Conversely, (2.37) implies (2.33). Thus (2.33) and (2.37) are equivalent. We have the following

Proposition 2.1.1 *$F = |y|\sqrt{f(\frac{|x|^2}{2}, \frac{\langle x,y\rangle}{|y|})}$ is a solution of the dually flat equation (2.1) if and only if f satisfies (2.37), where t and s are given in (2.6).*

2.1.3 Construction of Dually Flat Finsler Metrics

In this subsection we are going to give a lot of new dually flat Finsler metrics of orthogonal invariance, these metrics contains the Funk metric on the unit ball.

Consider the spherically symmetric Finsler metric $F = |y|\phi(\frac{|x|^2}{2}, \frac{\langle x,y\rangle}{|y|})$ on $\mathbb{B}^m(r_\mu)$ where $f = f(t, s)$ is given by

$$f(t, s) = \sum_{j=0}^{l} f_j(t)s^j \tag{2.38}$$

where f_j are differentiable functions. By simple calculations, we have

$$f_t(t, s) = \sum_{j=0}^{l} f_j'(t)s^j \tag{2.39}$$

and

$$f_{ts}(t, s) = \sum_{j=0}^{l} jf_j'(t)s^{j-1}. \tag{2.40}$$

Similarly, we have

$$f_{ss}(t, s) = \sum_{j=2}^{l} j(j-1)f_j(t)s^{j-2} = \sum_{k=0}^{l-2}(k+2)(k+1)f_{k+2}(t)s^k. \tag{2.41}$$

By using (2.39), (2.40) and (2.41) we get

$$sf_{ts} + f_{ss} - 2f_t = \sum_{j=0}^{l-2}\Big[(j-2)f_j' + (j+2)(j+1)f_{j+2}\Big]s^j + \sum_{j=l-1}^{l}(j-2)f_j's^j \tag{2.42}$$

where $f_j' = f_j'(t)$ and $f_{j+2} = f_{j+2}(t)$. By (2.42) and Proposition 2.1.1, $F = F(x, y)$ is dually flat if and only if

$$\begin{cases} (j-2)f_j'(t) = 0, & j = l-1,\ l, \\ (j-2)f_j'(t) + (j+2)(j+1)f_{j+2}(t) = 0, & j = 0, \cdots, l-2. \end{cases} \tag{2.43}$$

Taking $j = 0$ and using (2.43) we obtain

$$f_2(t) = f_0'(t). \tag{2.44}$$

Similarly, setting $j = 1$ and $j = 2$ in (2.43) respectively, we have

$$f_3(t) = \frac{1}{6} f_1'(t), \quad f_4(t) = 0. \tag{2.45}$$

It is easy to see that the second equation of (2.43) is equivalent to the following:

$$k(k-1) f_k(t) + (k-4) f_{k-2}'(t) = 0. \tag{2.46}$$

If $k =$ even ≥ 4, then

$$\begin{aligned} f_k(t) &= -\tfrac{k-4}{k(k-1)} f_{k-2}'(t) \\ &= (-1)^2 \tfrac{(k-4)(k-6)}{k(k-1)(k-2)(k-3)} f_{k-4}''(t) \\ &= \cdots\cdots = (-1)^{\frac{k-4}{2}} \tfrac{(k-4)(k-6)\cdots 4\times 2}{k(k-1)\cdots 7\times 6\times 5} f_4^{(\frac{k-4}{2})}(t) = 0 \end{aligned} \tag{2.47}$$

where we have used (2.46) and (2.45).

If $k =$ odd ≥ 5, then it follows from (2.46) and the first equation of (2.45) that

$$\begin{aligned} f_k(t) &= -\tfrac{k-4}{k(k-1)} f_{k-2}'(t) \\ &= \cdots\cdots = (-1)^{\frac{k-3}{2}} \tfrac{6(k-4)!!}{k!} f_3^{(\frac{k-3}{2})}(t) = (-1)^{\frac{k-3}{2}} \tfrac{(k-4)!!}{k!} f_1^{(\frac{k-1}{2})}(t). \end{aligned} \tag{2.48}$$

Case 1: $l =$ even ≥ 6. In this case, then

$$f_l(t) = 0, \qquad f_{l-1}(t) = \text{ constant}$$

where we have made use of (2.47) and (2.43). Set

$$l = 2n + 2, \quad g(t) = f_0(t), \quad h(t) = f_1(t).$$

It follows from (2.44), (2.45), (2.47) and (2.48) that

$$f(t,s) = g(t) + h(t)s + g'(t)s^2 + \frac{1}{6} h'(t)s^3 + \sum_{j=2}^{n} (-1)^{j-1} \frac{(2j-3)!!}{(2j+1)!} h^{(j)}(t) s^{2j+1} \tag{2.49}$$

and

$$h^{(n)}(t) = \text{ constant}. \tag{2.50}$$

Case 2: $l = \text{odd} \geq 5$. In this case,

$$f_l(t) = \text{ constant}, \qquad f_{l-1}(t) = \text{ constant } = 0.$$

Put

$$l = 2n + 1, \quad g(t) = f_0(t), \quad h(t) = f_1(t).$$

Then we also have (2.49) and (2.50).

The case $l \in \{1, 2, 3\}$ is similar. Thus we have the following:

Proposition 2.1.2 $F = |y|\sqrt{f(\frac{|x|^2}{2}, \frac{\langle x,y\rangle}{|y|})}$ *in the form* (2.2) *is a solution of the dually flat equation* (2.1) *if and only if* $f(t, s)$ *satisfies* (2.49) *and* (2.50).

Consider the solution f of (2.37) where $f = f(t, s)$ is given by

$$f(t, s) = \sqrt{\sum_{j=0}^{l} f_j(t)s^j}, \qquad f_l \neq 0. \tag{2.51}$$

It follows that

$$2ff_t = \sum_{j=0}^{l} f_j' s^j, \qquad 2ff_s = \sum_{j=0}^{l} jf_j s^{j-1} \tag{2.52}$$

and

$$2f_s f_t + 2ff_{ts} = \sum_{j=0}^{l} jf_j' s^{j-1}. \tag{2.53}$$

Combining (2.53) with (2.51) and (2.52), we get

$$\begin{aligned} 4f^3 f_{ts} &= 2f^2 \sum_{j=0}^{l} jf_j' s^{j-1} - 2ff_s \cdot 2ff_t \\ &= 2\sum_{i=0}^{l} f_i s^i \sum_{j=0}^{l} jf_j' s^{j-1} - \sum_{i=0}^{l} if_i s^{i-1} \sum_{j=0}^{l} f_j' s^j \\ &= \sum_{k=1}^{2l} \sum_{i+j=k} (2j - i) f_i f_j' s^{k-1} \end{aligned} \tag{2.54}$$

where we have made use of Lemma 2.1.3. Differentiating the second equation of (2.52) with respect to s, we obtain

$$2f_s^2 + 2ff_{ss} = \sum_{j=0}^{l} j(j-1)f_j s^{j-2}.$$

It follows from (2.51), the second equation of (2.52) and Lemma 2.1.3 that

$$\begin{aligned}4f^3 f_{ss} &= 2\sum_{i=0}^{l} f_i s^i \sum_{j=0}^{l} j(j-1)f_j s^{j-2} - \sum_{i=0}^{l} i f_i s^{i-1} \sum_{j=0}^{l} j f_j s^{j-1}\\ &= \sum_{k=2}^{2l} \sum_{i+j=k} j(2j-i-2) f_i f_j s^{k-2}\\ &= \sum_{l=0}^{2l-2} \sum_{i+j=l+2} j(2j-i-2) f_i f_j s^l \\ &= \sum_{k=0}^{2l-2} \sum_{p+q=k} (q+1)(2q-p-1) f_{p+1} f_{q+1} s^k\\ &= \sum_{k=0}^{2l-2} \sum_{i+j=k} (j+1)(2j-i-1) f_{i+1} f_{j+1} s^k.\end{aligned} \tag{2.55}$$

On the other hand, by using (2.51), the first equation of (2.52) and Lemma 2.1.3 we get

$$2f^3 f_t = \sum_{k=0}^{2l} \sum_{i+j=k} f_i f_j' s^k. \tag{2.56}$$

Together with (2.54) and (2.55) we have

$$\begin{aligned}&4f^3(sf_{ts} + f_{ss} - 2f_t)\\ &= \sum_{k=0}^{2l-2} \sum_{i+j=k} \Big[(2j-i-4) f_i f_j' + (j+1)(2j-i-1) f_{i+1} f_{j+1}\Big] s^k\\ &\quad + \sum_{k=2l-1}^{2l-2} \sum_{i+j=k} (2j-i-4) f_i f_j' s^k.\end{aligned} \tag{2.57}$$

It follows that f satisfies (2.37) if and only if

$$\begin{cases} \sum_{i+j=k} (2j-i-4) f_i f_j' = 0, \ k = 2l-1, \ 2l, \\ \sum_{i+j=k} \Big[(2j-i-4) f_i f_j' + (j+1)(2j-i-1) f_{i+1} f_{j+1}\Big] = 0, \\ k = 0, \ 1, \ \cdots, \ 2l-2. \end{cases} \tag{2.58}$$

Hence we have the following:

Proposition 2.1.3 $F = F(x, y)$ *in the form* (2.3) *is a solution of the dually flat equation* (2.1) *if and only if* $\{f_j\}$ *satisfy* (2.58).

Let us take a look at a special case, namely when $l = 4$ and $f_1(t) = f_3(t) = 0$, then

$$f_0(f_2 - f_0') = 0, \tag{2.59}$$

$$4f_0f_4 - f_2f_0' = 0, \tag{2.60}$$

$$6f_2f_4 - 4f_4f_0' - f_2f_2' + 2f_0f_4' = 0, \tag{2.61}$$

$$4f_4^2 - 2f_4f_2' + f_2f_4' = 0. \tag{2.62}$$

Case 1: $f_0 = 0$. In this case, (2.59) and (2.60) hold automatically. Using (2.61), we have

$$f_2(6f_4 - f_2') = 0. \tag{2.63}$$

If $f_2 \neq 0$, then

$$f_2' = 6f_4 \tag{2.64}$$

where we have used (2.63). Plugging this into (2.62) yields $f_2f_4' = 8f_4^2 = \frac{4}{3}f_4f_2'$. Solving it, we get

$$f_4 = cf_2^{\frac{4}{3}}, \qquad c = constant. \tag{2.65}$$

Together with (2.64) we have $\frac{df_2}{dt} = 6cf_2^{\frac{4}{3}}$. It follows that

$$f_2(t) = \frac{c_2}{(c_1 + t)^3}. \tag{2.66}$$

Together with (2.64), we conclude that

$$f_4(t) = \frac{1}{6}f_2'(t) = -\frac{c_2}{2(c_1 + t)^4}. \tag{2.67}$$

If $f_2 = 0$, then it follows from (2.62) that $f_4 = 0$. Hence (2.66) and (2.67) also hold.

Case 2: $f_0 \neq 0$. In this case, then

$$f_2 = f_0' \tag{2.68}$$

where we have made use of (2.59). Substituting (2.68) into (2.60) yields

$$f_4 = \frac{(f_0')^2}{4f_0}. \tag{2.69}$$

By (2.68) and (2.69), we obtain that (2.61) and (2.62) are identical relations. Thus we have the following:

Proposition 2.1.4 *We have the following solutions of* (2.1)*:*

$$\text{(i) } F = |y|\sqrt{f\left(\frac{|x|^2}{2}, \frac{\langle x, y\rangle}{|y|}\right)}, \quad f(t, s) = \sqrt{\frac{c_2 s^2}{(c+t)^3} - \frac{c_2 s^4}{2(c+t)^4}};$$

$$\text{(ii) } F = |y|\sqrt{f\left(\frac{|x|^2}{2}, \frac{\langle x, y\rangle}{|y|}\right)}, \quad f(t, s) = \sqrt{h(t) + h'(t)s^2 + \frac{(h'(t))^2}{4h(t)}s^4}$$

where h is an any differentiable function.

Proof of Theorem 2.1.1 Using Propositions 2.1.2, 2.1.4 (i) and the fundamental property of the dually flat equation (2.1).

Similarly we have the following:

Theorem 2.1.2 *Let $f(t, s)$ be a function defined by*

$$f(t, s) = g(t) + h(t)s + g'(t)s^2 + \tfrac{1}{6}h'(t)s^3 + \sum_{j=2}^{n}(-1)^{j-1}\tfrac{(2j-3)!!}{(2j+1)!}h^{(j)}(t)s^{2j+1}$$
$$+b\sqrt{\lambda(t) + \lambda'(t)s^2 + \tfrac{(\lambda'(t))^2}{4\lambda(t)}s^4}$$

where b is a constant and g, h and λ are any differentiable functions. Then the following spherically symmetric Finsler metric on $\mathbb{B}^m(r_\mu)$ given in (2.5) *is dually flat.*

2.2 Dually Flat Metrics and Pogorelov' Integral Representation

Proposition 2.1.1 tells us the following interesting fact: the dual flatness for a spherically symmetric Finsler metric is independent of the dimension of metric. It follows that in order to construct dually flat spherically symmetric Finsler metrics we can only consider the two-dimensional case.

2.2.1 Two-Dimensional Hamel' Differential Equation

We know that a two-dimensional Finsler metric $F = F(x, y)$ on an open subset $\mathcal{U} \subset \mathbb{R}^2$ is projectively flat if an only if F satisfies the following two-dimensional Hamel's equations [23]

$$\frac{\partial^2 F}{\partial x^2 \partial y^1} = \frac{\partial^2 F}{\partial x^1 \partial y^2}. \tag{2.70}$$

In this subsection we are going to construct solutions of two-dimensional Hamel's equations (2.70) generalizing result previously only known in the case of reversible two-dimensional Finsler metrics.

A Finsler metric $\Theta = \Theta(x, y)$ on an open subset $\mathcal{U} \subset \mathbb{R}^K$ is said to be *projectively flat* if all geodesics are straight in $\mathcal{U}$, equivalently, it satisfies the following system of equations [23],

$$\Theta_{x^j y^i} y^j = \Theta_{x^i} \tag{2.71}$$

where $x = (x^1, \cdots, x^K) \in \mathcal{U}$ and $y = y^j \frac{\partial}{\partial x^j}|_x \in T_x\mathcal{U}$. We call (2.71) the *projectively flat equation*.

Lemma 2.2.1 ([23]) *Assume $\Theta : T\mathcal{U} \to \mathbb{R}$ is positively homogeneous of degree one with respect to y. Then Θ is a solution of (2.71) if and only if it satisfies the following system of equations*

$$\Theta_{x^i y^j} = \Theta_{x^j y^i}. \tag{2.72}$$

Lemma 2.2.2 *Define*

$$\Theta(x, y) := \int_{\phi-\frac{\pi}{2}}^{\phi+\frac{\pi}{2}} (y^1 \cos\theta + y^2 \sin\theta) f(x, \theta) d\theta \tag{2.73}$$

where f is a function on $\mathcal{U} \times \mathbb{R}$, $y = (y^1, y^2)$ and

$$\phi = \arg(y^1 + \sqrt{-1} y^2) \tag{2.74}$$

i.e. ϕ is the argument of the complex number $y^1 + \sqrt{-1}y^2$. Then

$$\frac{\partial \Theta}{\partial y^j} = \int_{\phi-\frac{\pi}{2}}^{\phi+\frac{\pi}{2}} \frac{\partial}{\partial y^j} \Big[(y^1 \cos\theta + y^2 \sin\theta) f(x, \theta)\Big] d\theta$$

where $j \in \{1, 2\}$.

Proof By using (2.74), we can express $y = (y^1, y^2)$ in the polar coordinate system,

$$y^1 = r\cos\phi, \qquad y^2 = r\sin\phi. \tag{2.75}$$

Hence $\Theta = \Theta(x, y)$ is given by

$$\Theta(x, y) = \int_{\phi-\frac{\pi}{2}}^{\phi+\frac{\pi}{2}} r(\cos\phi\cos\theta + \sin\phi\sin\theta) f(x, \theta)d\theta = rh(x, \phi) \tag{2.76}$$

where

$$h(x, \phi) := \int_{\phi-\frac{\pi}{2}}^{\phi+\frac{\pi}{2}} f(x, \theta)\cos(\phi - \theta)d\theta. \tag{2.77}$$

It follows that (2.77) is the cosine transform of f for each fixed x. Observe that by (2.75),

$$\frac{\partial}{\partial y^1} = \cos\phi\frac{\partial}{\partial r} - \frac{1}{r}\sin\phi\frac{\partial}{\partial\phi}, \qquad \frac{\partial}{\partial y^2} = \sin\phi\frac{\partial}{\partial r} + \frac{1}{r}\cos\phi\frac{\partial}{\partial\phi}. \tag{2.78}$$

Together with (2.76) we have

$$\frac{\partial\Theta}{\partial y^1} = (\cos\phi)\frac{\partial\Theta}{\partial r} - \frac{1}{r}(\sin\phi)\frac{\partial\Theta}{\partial\phi} = h\cos\phi - \frac{\partial h}{\partial\phi}\sin\phi. \tag{2.79}$$

Similarly, we get

$$\frac{\partial\Theta}{\partial y^2} = h\sin\phi + \frac{\partial h}{\partial\phi}\cos\phi. \tag{2.80}$$

Now we compute $\frac{\partial h}{\partial\phi}$ in (2.79). Let

$$g(x, \phi, \theta) = \int f(x, \theta)\cos(\phi - \theta)d\theta. \tag{2.81}$$

It follows that

$$\frac{\partial g}{\partial\theta} = f(x, \theta)\cos(\phi - \theta) \tag{2.82}$$

from which together with (2.77) we obtain

$$h(x, \phi) = g(x, \phi, \theta)|_{\theta=\phi-\frac{\pi}{2}}^{\theta=\phi+\frac{\pi}{2}} = g\left(x, \phi, \phi + \frac{\pi}{2}\right) - g\left(x, \phi, \phi - \frac{\pi}{2}\right). \tag{2.83}$$

Thus we have

$$\begin{aligned}\frac{\partial h}{\partial \phi} &= \frac{\partial g}{\partial \phi}\left(x, \phi, \phi+\frac{\pi}{2}\right)+\frac{\partial g}{\partial \theta}\left(x, \phi, \phi+\frac{\pi}{2}\right)-\frac{\partial g}{\partial \phi}\left(x, \phi, \phi-\frac{\pi}{2}\right)\\ &\quad -\frac{\partial g}{\partial \theta}\left(x, \phi, \phi-\frac{\pi}{2}\right)\\ &= \left(\frac{\partial g}{\partial \phi}+\frac{\partial g}{\partial \theta}\right)(x, \phi, \theta)\Big|_{\theta=\phi-\frac{\pi}{2}}^{\theta=\phi+\frac{\pi}{2}}.\end{aligned} \tag{2.84}$$

By (2.81), we have

$$\frac{\partial g}{\partial \phi}=\int f(x, \theta)\left[\frac{\partial}{\partial \phi}\cos(\phi-\theta)\right]d\theta=-\int f(x, \theta)\sin(\phi-\theta)d\theta. \tag{2.85}$$

By using (2.82), we get

$$\frac{\partial g(x, \phi, \theta)}{\partial \theta}\Big|_{\theta=\phi-\frac{\pi}{2}}^{\theta=\phi+\frac{\pi}{2}}=0. \tag{2.86}$$

Together with (2.84) and (2.85) yields

$$\begin{aligned}\frac{\partial h}{\partial \phi} &= \frac{g(x, \phi, \theta)}{\partial \phi}\Big|_{\theta=\phi-\frac{\pi}{2}}^{\theta=\phi+\frac{\pi}{2}}\\ &= -\left[\int f(x, \theta)\sin(\phi-\theta)d\theta\right]_{\theta=\phi-\frac{\pi}{2}}^{\theta=\phi+\frac{\pi}{2}}\\ &= -\int_{\phi-\frac{\pi}{2}}^{\phi+\frac{\pi}{2}} f(x, \theta)\sin(\phi-\theta)d\theta.\end{aligned} \tag{2.87}$$

Plugging (2.77) and (2.87) into (2.79) yields

$$\begin{aligned}\frac{\partial \Theta}{\partial y^1} &= (\cos\phi)\int_{\phi-\frac{\pi}{2}}^{\phi+\frac{\pi}{2}} f(x, \theta)\cos(\phi-\theta)d\theta+(\sin\phi)\int_{\phi-\frac{\pi}{2}}^{\phi+\frac{\pi}{2}} f(x, \theta)\sin(\phi-\theta)d\theta\\ &= \int_{\phi-\frac{\pi}{2}}^{\phi+\frac{\pi}{2}} f(x, \theta)\left[\cos\phi\cos(\phi-\theta)+\sin\phi\sin(\phi-\theta)\right]d\theta\\ &= \int_{\phi-\frac{\pi}{2}}^{\phi+\frac{\pi}{2}} f(x, \theta)\cos\theta d\theta=\int_{\phi-\frac{\pi}{2}}^{\phi+\frac{\pi}{2}}\frac{\partial}{\partial y^1}\left[(y^1\cos\theta+y^2\sin\theta)f(x, \theta)\right]d\theta.\end{aligned} \tag{2.88}$$

Similarly, we have

$$\begin{aligned}\frac{\partial \Theta}{\partial y^2} &= \int_{\phi-\frac{\pi}{2}}^{\phi+\frac{\pi}{2}} f(x, \theta)\left[\sin\phi\cos(\phi-\theta)+\cos\phi\sin(\phi-\theta)\right]d\theta\\ &= \int_{\phi-\frac{\pi}{2}}^{\phi+\frac{\pi}{2}}\frac{\partial}{\partial y^2}\left[(y^1\cos\theta+y^2\sin\theta)f(x, \theta)\right]d\theta.\end{aligned} \tag{2.89}$$

Thus we complete the proof of Lemma 2.2.2.

Corollary 2.2.1 *Let $\Theta = \Theta(x, y)$ denote the function on $T\mathcal{U}$ defined by* (2.73). *The function Θ satisfies Hamel's differential equation if and only if*

$$\int_{\phi-\frac{\pi}{2}}^{\phi+\frac{\pi}{2}} \left[(\cos\theta)\frac{\partial f}{\partial x^2} - (\sin\theta)\frac{\partial f}{\partial x^1}\right] d\theta = 0 \tag{2.90}$$

for any ϕ.

Proof Recall that in the two-dimensional case, the Hamel's PDE is given in (2.70). Lemma 2.2.2 gives that

$$\begin{aligned}
\frac{\partial^2\Theta}{\partial x^2\partial y^1} - \frac{\partial^2\Theta}{\partial x^1\partial y^2} &= \frac{\partial}{\partial x^2}\left(\frac{\partial\Theta}{\partial y^1}\right) - \frac{\partial}{\partial x^1}\left(\frac{\partial\Theta}{\partial y^2}\right)\\
&= \frac{\partial}{\partial x^2}\int_{\phi-\frac{\pi}{2}}^{\phi+\frac{\pi}{2}} f(x, \theta)\cos\theta d\theta - \frac{\partial}{\partial x^1}\int_{\phi-\frac{\pi}{2}}^{\phi+\frac{\pi}{2}} f(x, \theta)\sin\theta d\theta\\
&= \int_{\phi-\frac{\pi}{2}}^{\phi+\frac{\pi}{2}} \left[(\cos\theta)\frac{\partial f}{\partial x^2} - (\sin\theta)\frac{\partial f}{\partial x^1}\right] d\theta.
\end{aligned}$$

Thus (2.70) is equivalent to (2.90).

Proposition 2.2.1 *Let $\Theta : T\mathcal{U} \to \mathbb{R}$ be a function defined by*

$$\Theta(x^1, x^2; y^1, y^2) = \int_{\phi-\frac{\pi}{2}}^{\phi+\frac{\pi}{2}} (y^1\cos\theta + y^2\sin\theta)\rho(x^1\cos\theta + x^2\sin\theta, \theta)d\theta \tag{2.91}$$

where ϕ is the argument of the complex number $y^1 + \sqrt{-1}y^2$. Then the function Θ satisfies Hamel's differential equation (2.70).

Proof Let us consider the solution of the following linear partial differential equation

$$(\cos\theta)\frac{\partial f}{\partial x^2} - (\sin\theta)\frac{\partial f}{\partial x^1} = 0. \tag{2.92}$$

The characteristic equation of linear PDE (2.92) is

$$\frac{dx^1}{\sin\theta} = \frac{dx^2}{\cos\theta} = \frac{d\theta}{0}. \tag{2.93}$$

It follows that $x^1\cos\theta + x^2\sin\theta = c_1$, $\theta = c_2$ are independent integrals of (2.93). Therefore the solution of (2.92) is

$$f(x^1, x^2, \theta) = \rho(x^1\cos\theta + x^2\sin\theta, \theta) \tag{2.94}$$

where $\rho(\,,\,)$ is a continuously differentiable function. It follows that Θ satisfies Hamel's differential equation (2.70) from Corollary 2.2.1.

Remark 2.2.5 Using the Semicircle transformation, it is easy to show that all reversible solutions of two-dimensional Hamel's differential equation (2.70) are given by (2.91) [1, 61].

2.2.2 A New Approach to Produce Solutions of Dually Flat Equations

In this subsection we are going to present a new approach to produce two-dimensional solutions of dually flat equation (2.1).

Lemma 2.2.3 *Let $\mathscr{U}$ be an open subset in $\mathbb{R}^K$. Assume that $F : T\mathscr{U} \to \mathbb{R}$ is positively homogenous of degree one with respect to y. Then F is a solution of (2.1) if and only if it satisfies the following equations:*

$$L_{x^i y^j} = L_{x^j y^i} \tag{2.95}$$

where $L := F^2/2$.

Proof A function ξ defined on $T\mathscr{U}$ can be expressed as $\xi(x^1, \cdots, x^K; y^1, \cdots, y^K)$. We use the following notation $\xi_0 := \frac{\partial \xi}{\partial x^i} y^i$. Note that L is positively homogeneous of degree two. Hence L_{x^i} is also positively homogeneous of degree two, i.e. $L_{x^i}(x, \lambda y) = \lambda^2 L_{x^i}(x, y)$. It follows that

$$L_{x^i y^j} y^j = 2L_{x^i}. \tag{2.96}$$

First suppose that F satisfies

$$L_{x^j y^i} y^j = 2L_{x^i}. \tag{2.97}$$

On the other hand,

$$(L_0)_{y^i} = (L_{x^j} y^j)_{y^i} = L_{x^j y^i} y^j + L_{x^i}. \tag{2.98}$$

Combining this with (2.97), we get $2L_{x^i} = (L_0)_{y^i} - L_{x^i}$, that is

$$(L_0)_{y^i} = 3L_{x^i}. \tag{2.99}$$

Differentiating (2.99) with respect to y^j, we obtain $L_{x^i y^j} = \frac{1}{3}(L_0)_{y^i y^j} = \frac{1}{3}(L_0)_{y^j y^i} = L_{x^j y^i}$. Thus we obtain (2.95).

Conversely, suppose that (2.95) holds. Together with (2.96) we have (2.97), i.e. (2.1) holds.

Proposition 2.2.2 *Let $\Theta : T\mathscr{U} \to \mathbb{R}$ satisfies partial differential equation* (2.72). *Then*

$$F := \sqrt{\Theta_{x^j} y^j} \tag{2.100}$$

is a solution of the dually flat equation (2.1).

Proof Suppose $\Theta = \Theta(x, y)$ satisfies (2.72), then it satisfies (2.71) from Lemma 2.2.1. Differentiating (2.100) with respect to x^j, we have $L_{x^j} = \left(\frac{F^2}{2}\right)_{x^j} = \frac{1}{2}\Theta_{x^i x^j} y^i$. It follows that

$$L_{x^j y^k} = \frac{1}{2}(I) + \frac{1}{2}\Theta_{x^k x^j} \tag{2.101}$$

where

$$(I) := \Theta_{x^i x^j y^k} y^i = \Theta_{x^i y^k x^j} y^i = (\Theta_{x^i y^k} y^i)_{x^j} = \Theta_{x^k x^j} \tag{2.102}$$

where we have used (2.71). Plugging (2.102) into (2.101) yields $L_{x^j y^k} = \Theta_{x^k x^j}$. Note that $\Theta_{x^j x^k} = \Theta_{x^k x^j}$. Hence we obtain (2.95). By Lemma 2.2.3, F is a solution of the dually flat equation (2.1).

Theorem 2.2.1 *Let $F : T\mathscr{U} \to \mathbb{R}$ be a function defined by*

$$F(x^1, x^2; y^1, y^2) = \left\{\int_{\phi-\frac{\pi}{2}}^{\phi+\frac{\pi}{2}} (y^1 \cos\theta + y^2 \sin\theta)^2 \zeta(x^1 \cos\theta + x^2 \sin\theta, \theta) d\theta\right\}^{\frac{1}{2}} \tag{2.103}$$

where $\zeta(,)$ is a positive continuous function and ϕ is the argument of the complex number $y^1 + \sqrt{-1}y^2$. Then the function F satisfies the dually flat equation (2.1).

Proof We consider the function Θ defined in (2.91). Then

$$\frac{\partial \Theta}{\partial x^1} = \int_{\phi-\frac{\pi}{2}}^{\phi+\frac{\pi}{2}} (y^1 \cos\theta + y^2 \sin\theta)\zeta(s, \theta) \cos\theta d\theta \tag{2.104}$$

and

$$\frac{\partial \Theta}{\partial x^2} = \int_{\phi-\frac{\pi}{2}}^{\phi+\frac{\pi}{2}} (y^1 \cos\theta + y^2 \sin\theta)\zeta(s, \theta) \sin\theta d\theta \tag{2.105}$$

where $\zeta(s, \theta) = \frac{\partial \rho(s, \theta)}{\partial s}$. Plugging (2.104) and (2.105) into (2.100) yields (2.103). Combining this with Propositions 2.2.1 and 2.2.2, we have Theorem 2.2.1.

2.2.3 New Dually Flat Spherically Symmetric Metrics

In this subsection, we are going to prove Theorems 2.2.2 and 2.2.3 below. In (2.91) we take $\rho(s, \theta) = s^m$ where $m \in \{1, 2, \cdots\}$. Then we obtain

$$\rho(x^1 \cos\theta + x^2 \sin\theta,\ \theta) = (x^1 \cos\theta + x^2 \sin\theta)^m. \tag{2.106}$$

By a straightforward computation one obtains

$$\rho(x^1 \cos\theta + x^2 \sin\theta,\ \theta) = R^m \cos^m(\theta - \psi) \tag{2.107}$$

where

$$x^1 = R\cos\psi, \qquad x^2 = R\sin\psi. \tag{2.108}$$

From (2.75), (2.91) and (2.107) we have

$$\begin{aligned}\Theta(x, y) &= rR^m \int_{\phi-\frac{\pi}{2}}^{\phi+\frac{\pi}{2}} \cos(\theta - \phi)\cos^m(\theta - \psi)d\theta \\ &= rR^m \int_{-\frac{\pi}{2}}^{\frac{\pi}{2}} \cos\alpha \cos^m(\alpha + \beta)d\alpha\end{aligned} \tag{2.109}$$

where

$$\alpha := \theta - \phi, \qquad \beta := \theta - \psi. \tag{2.110}$$

Case 1: $m = 2n$, where $n \in \{1, 2, \cdots\}$.

Observe that

$$\cos^{2n}(\alpha + \beta) = \frac{1}{2^{2n-1}}\left[\frac{1}{2}C_{2n}^n + \sum_{k=0}^{n-1} C_{2n}^k \cos(2n - 2k)(\alpha + \beta)\right].$$

It follows that

$$\begin{aligned}\cos\alpha\cos^{2n}(\alpha+\beta) &= \tfrac{1}{2^{2n}}C_{2n}^n\cos\alpha + \tfrac{1}{2^{2n}}\sum_{k=0}^{n-1} C_{2n}^k\,[\cos(2n-2k+1)\alpha + 2(n-k)\beta] \\ &\quad + \tfrac{1}{2^{2n}}\sum_{k=0}^{n-1} C_{2n}^k\,[\cos(2n-2k-1)\alpha + 2(n-k)\beta].\end{aligned} \tag{2.111}$$

By simple calculations, we have

$$\int_{-\frac{\pi}{2}}^{\frac{\pi}{2}} \cos\alpha d\alpha = 2, \tag{2.112}$$

$$\int_{-\frac{\pi}{2}}^{\frac{\pi}{2}} \cos[(2n-2k+1)\alpha + 2(n-k)\beta]\, d\alpha = \frac{(-1)^{n-k}2}{2(n-k)+1} \cos[2(n-k)\beta], \tag{2.113}$$

$$\int_{-\frac{\pi}{2}}^{\frac{\pi}{2}} \cos[(2n-2k-1)\alpha + 2(n-k)\beta]\, d\alpha = \frac{(-1)^{n-k}2}{2(n-k)-1} \cos[2(n-k)\beta]. \tag{2.114}$$

Plugging (2.111) into (2.109) and using (2.112), (2.113) and (2.114) we get

$$\begin{aligned} \Theta(x, y) &= \tfrac{rR^{2n}}{2^{2n}} C_{2n}^n \int_{-\frac{\pi}{2}}^{\frac{\pi}{2}} \cos\alpha d\alpha \\ &\quad + \tfrac{rR^{2n}}{2^{2n}} \sum_{k=0}^{n-1} C_{2n}^k \int_{-\frac{\pi}{2}}^{\frac{\pi}{2}} \cos[(2n-2k+1)\alpha + 2(n-k)\beta]\, d\alpha \\ &\quad + \tfrac{rR^{2n}}{2^{2n}} \sum_{k=0}^{n-1} C_{2n}^k \int_{-\frac{\pi}{2}}^{\frac{\pi}{2}} \cos[(2n-2k-1)\alpha + 2(n-k)\beta]\, d\alpha \\ &= \tfrac{rR^{2n}}{2^{2n-1}} \left\{ C_{2n}^n + \sum_{k=0}^{n-1} \tfrac{(-1)^{n-k+1} C_{2n}^k}{4(n-k)^2-1} 2\cos[2(n-k)\beta] \right\}. \end{aligned} \tag{2.115}$$

A direct computation gives

$$\cos[2(n-k)\beta] = \sum_{l=0}^{n-k} (-1)^l C_{2(n-k)}^{2l} \cos^{2(n-k-l)}\beta \left(1-\cos^2\beta\right)^l. \tag{2.116}$$

Substituting (2.116) into (2.115) yields

$$\Theta(x, y) = \frac{rR^{2n}}{2^{2n-1}} \left\{ C_{2n}^n + 2\sum_{k=0}^{n-1}\sum_{l=0}^{n-k} a(n, k, l) \cos^{2(n-k-l)}\beta \left(1-\cos^2\beta\right)^l \right\} \tag{2.117}$$

where

$$a(n, k, l) := \frac{(-1)^{n-k+1} C_{2n}^k C_{2(n-k)}^{2l}}{4(n-k)^2-1}. \tag{2.118}$$

Using (2.75), (2.108) and (2.110) we have

$$r = |y|, \quad R = |x|, \quad \cos\beta = \frac{\langle x, y\rangle}{|x||y|} \tag{2.119}$$

where $|\cdot|$ and $\langle\cdot\rangle$ denote the standard Euclidean norm and inner product in $\mathbb{R}^2$. Now we compute $\Theta_0 := \Theta_{x^j} y^j$. First, observe that

$$\left(R^{2n}\right)_0 = n\left(R^2\right)^{n-1}\left(R^2\right)_0 = 2nR^{2n-2}\langle x, y\rangle, \tag{2.120}$$

$$(\cos\beta)_0 = \frac{1}{|y|}\left(\frac{\langle x,\ y\rangle}{|x|}\right)_0 = \frac{|y|}{|x|}\left(1 - \frac{\langle x,\ y\rangle^2}{|x|^2|y|^2}\right) = \frac{r}{R}\sin^2\beta. \tag{2.121}$$

By (2.117), we obtain

$$\begin{aligned}\Theta_0 = &\tfrac{r}{2^{2n-1}}\left(R^{2n}\right)_0\left[C_{2n}^n + 2\sum_{k=0}^{n-1}\sum_{l=0}^{n-k} a(n,\ k,\ l)\cos^{2(n-k-l)}\beta\sin^{2l}\beta\right]\\ &+\tfrac{rR^{2n}}{2^{2n-3}}\sum_{k=0}^{n-1}\sum_{l=0}^{n-k}(I)_{kl}a(n,\ k,\ l)\,(\cos\beta)_0\end{aligned} \tag{2.122}$$

where

$$\begin{aligned}(I)_{kl} &= (n-l-k)\cos^{2(n-k-l)-1}\beta\sin^{2l}\beta - l\cos^{2(n-k-l)+1}\beta\sin^{2l-2}\beta\\ &= -\left[l-(n-k)\sin^2\beta\right]\cos^{2(n-k-l)-1}\beta\sin^{2l-2}\beta.\end{aligned} \tag{2.123}$$

Substituting (2.120), (2.121) and (2.123) into (2.122) we have

$$\begin{aligned}\Theta_0 = &\tfrac{nr}{2^{2n-2}}R^{2n-2}\langle x,\ y\rangle\left[C_{2n}^n + 2\sum_{k=0}^{n-1}\sum_{l=0}^{n-k} a(n,\ k,\ l)\cos^{2(n-k-l)}\beta\sin^{2l}\beta\right]\\ &+\tfrac{r^2R^{2n-1}}{2^{2n-3}}\sum_{k=0}^{n-1}\sum_{l=0}^{n-k} a(n,\ k,\ l)\left[(n-k)\sin^2\beta - l\right]\cos^{2(n-k-l)+1}\beta\sin^{2l}\beta\\ = &\tfrac{r^2R^{2n-1}}{2^{2n-2}}C_{2n}^n\cos\beta\\ &+\tfrac{r^2R^{2n-1}}{2^{2n-3}}\sum_{k=0}^{n-1}\sum_{l=0}^{n-k} a(n,\ k,\ l)(n-l-k\sin^2\beta)\cos^{2(n-k-l)-1}\beta\sin^{2l}\beta.\end{aligned}$$

Combining with Propositions 2.2.1, 2.2.2 and 2.1.1, we have the following

Theorem 2.2.2 *On $\mathbb{B}^K(r_\mu)$, the orthogonally invariant Finsler metric F defined by*

$$\begin{aligned}F^2 = &\epsilon|y|^2 + \tfrac{|x|^{2n-1}|y|^2}{2^{2n-2}}\times\\ &\times\left[C_{2n}^n\cos\beta + 2\sum_{k=0}^{n-1}\sum_{l=0}^{n-k} a(n,\ k,\ l)(n-l-k\sin^2\beta)\cos^{2(n-k-l)-1}\beta\sin^{2l}\beta\right]\end{aligned}$$

is dually flat where $n \in \{1,\ 2,\ \cdots\}$ and $a(n,\ k,\ l)$ and $\cos\beta$ are defined in (2.118) *and* (2.119) *respectively.*

Case 2: $m = 2n+1$, where $n \in \{1,\ 2,\ \cdots\}$.

$$\cos^{2n+1}(\alpha+\beta) = \frac{1}{2^{2n}}\sum_{k=0}^{n} C_{2n+1}^k\cos(2n-2k+1)(\alpha+\beta). \tag{2.124}$$

It follows that

$$\cos\alpha\cos^{2n+1}(\alpha+\beta) = \tfrac{1}{2^{2n}}\sum_{k=0}^{n} C_{2n+1}^{k}\cos\left[2(n-k+1)\alpha+(2n-2k+1)\beta\right] \\ +\tfrac{1}{2^{2n}}\sum_{k=0}^{n} C_{2n+1}^{k}\cos\left[2(n-k)\alpha+(2n-2k+1)\beta\right]. \tag{2.125}$$

By a straightforward computation one obtains

$$\int_{-\frac{\pi}{2}}^{\frac{\pi}{2}} \cos\left[2(n-k+1)\alpha+(2n-2k+1)\beta\right]d\alpha = 0. \tag{2.126}$$

Similarly, we get

$$\int_{-\frac{\pi}{2}}^{\frac{\pi}{2}} \cos\left[2(n-k)\alpha+(2n-2k+1)\beta\right]d\alpha = \begin{cases} 0, & if\ k<n, \\ \pi\cos\beta, & if\ k=n. \end{cases} \tag{2.127}$$

Substituting (2.125) into (2.109) and using (2.126)–(2.127) we obtain

$$\begin{aligned}\Theta(x,\,y) &= \frac{rR^{2n+1}}{2^{2n}}\sum_{k=0}^{n} C_{2n+1}^{k}\int_{-\frac{\pi}{2}}^{\frac{\pi}{2}} \cos\left[2(n-k+1)\alpha+(2n-2k+1)\beta d\alpha\right] \\ &\quad +\frac{rR^{2n+1}}{2^{2n}}\sum_{k=0}^{n} C_{2n+1}^{k}\int_{-\frac{\pi}{2}}^{\frac{\pi}{2}} \cos\left[2(n-k)\alpha+(2n-2k+1)\beta d\alpha\right] \\ &= rR^{2n+1}\tfrac{C_{2n+1}^{n}}{2^{2n}}\pi\cos\beta \end{aligned} \tag{2.128}$$

where r, R and $\cos\beta$ are defined in (2.119). Now we compute Θ_0. First, observe that

$$\left(R^{2n+1}\right)_0 = (2n+1)R^{2n}R_0 = (2n+1)R^{2n-1}\langle x,\,y\rangle. \tag{2.129}$$

By using (2.121), (2.128) and (2.129), we have

$$\begin{aligned}\Theta_0 &= \frac{\pi C_{2n+1}^{n}}{2^{2n}}r\left[\left(R^{2n+1}\right)_0\cos\beta + R^{2n+1}\left(\cos\beta\right)_0\right] \\ &= \frac{\pi C_{2n+1}^{n}}{2^{2n}}r^2R^{2n}\left(1+2n\cos^2\beta\right).\end{aligned}$$

Combining this with Propositions 2.2.1 and 2.2.2 we have the following:

Theorem 2.2.3 *The following orthogonally invariant Finsler metric*

$$F := |y|\left[\epsilon + \frac{\pi C_{2n+1}^{n}}{2^{2n}}|x|^{2n}\left(1+2n\cos^2\beta\right)\right]^{\frac{1}{2}}$$

is dually flat where $a(n,\,k,\,l)$ *and* $\cos\beta$ *are given in* (2.118) *and* (2.119) *respectively.*

2.3 Dually Flat Metrics and Projectively Flat Metrics

In this section we show that any solution of dually flat equations produces a solution of Hamel equations and vice versa (see Theorem 2.3.2). Using this correspondence, we are able to manufacture new dually flat Finsler metrics from known projectively flat Finsler metrics.

More precisely, we investigate how to construct the solutions of dually flat equations (2.1) from a projective spherically symmetric Finsler metric and seek conditions of producing Finsler metrics.

Recall that a Finsler metric $F = F(x, y)$ is called to be *spherically symmetric* if F satisfies (1.1) for all $A \in O(n)$, equivalently, the orthogonal group $O(n)$ act as isometries of F. Proposition 1.3.1 proves that any spherically symmetric Finsler metric $F = F(x, y)$ can be expressed by

$$F(x, y) = |y|\psi\left(|x|, \frac{\langle x, y\rangle}{|y|}\right).$$

First, we give an explicit expression of the solution of dually flat equations (2.1) corresponding a projectively flat Finsler metric (see Proposition 2.3.1 below). Next, we produce many new spherically symmetric dually flat Finsler metric by using Huang-Mo metrics in Proposition 2.3.2. More precisely, we prove the following:

Theorem 2.3.1 *Let $f(\lambda)$ be a polynomial function defined by*

$$f(\lambda) = 1 + \delta\lambda + 2n\Sigma_{k=0}^{n-1}\frac{(-1)^k C_{n-1}^k \lambda^{2k+2}}{(2k+1)(2k+2)} \tag{2.130}$$

where $C_m^k = \frac{m(m-1)\cdots(m-k+1)}{k!}$. Suppose that $f(-1) < 0$. Then the following Finsler metric on an open subset in $\mathbb{R}^n\backslash\{0\}$

$$F = |y|\left\{|x|^{2n-1}\left[2n\lambda f(\lambda) + (1-\lambda^2)f'(\lambda)\right]\right\}^{\frac{1}{2}}$$

is dually flat where $\lambda = \frac{\langle x,y\rangle}{|x||y|}$.

We have the following two interesting special cases:

(a) When $n = 1$, then

$$F = \frac{\sqrt{\delta\langle x, y\rangle^2 + 4|x||y|\langle x, y\rangle + \delta|x|^2|y|^2}}{|x|^{\frac{1}{2}}}$$

is dually flat where $\delta > 2$.

(b) When $n = 2$, then

$$F = \frac{\sqrt{\delta|x|^3|y|^3 + 8|x|^2|y|^2\langle x, y\rangle + 3\delta|x||y|\langle x, y\rangle^2 + \frac{8}{3}\langle x, y\rangle^3}}{|y|^{\frac{1}{2}}}$$

is dually flat where $\delta > \frac{8}{3}$.

Finally we should point out that the notions of dual flat and projectively flat are not equivalent. For example, the Finsler metric F in Example 1.2.4 is projectively flat [37], but F is not dually flat. This fact follows from Cheng-Shen-Zhou's Proposition 2.6 in [14] (if a Finsler metric is dually flat and projectively flat, then it is of constant flag curvature) and the classification theorem of projective spherically symmetric Finsler metrics of constant flag curvature due to L.Zhou and Mo-Zhu (see Sect. 7.1 below). Very recently, C.Yu has constructed the following new dually flat Finsler metrics [81]

$$F(x, y) = \left(1 + |x|^2\right)^{\frac{1}{4}} |y| \pm \left(1 + |x|^2\right)^{-\frac{1}{4}} \langle x, y\rangle.$$

Based on the above arguments, we obtain F is not projectively flat.

2.3.1 Dually Flat Equations

In this subsection we are going to explore some nice properties of dually flat equations. In particular, we show any solution of Hamel equations produces a solution of dually flat equations (see Theorem 2.3.2 below).

Lemma 2.3.1 *If $F : T\mathscr{U} \to \mathbb{R}$ is a solution of* (2.1) *where $\mathscr{U}$ is an open subset in $\mathbb{R}^n$, then there exists a function Θ such that*

$$\Theta_{x^i} = F F_{y^i}. \tag{2.131}$$

Proof Let

$$p_i = \left(\frac{F^2}{2}\right)_{y^i} = L_{y^i}. \tag{2.132}$$

Using (2.95), one obtains

$$(p_j)_{x^i} = L_{y^j x^i} = L_{y^i x^j} = (p_i)_{x^j}. \tag{2.133}$$

Take fixed $x_0 \in \mathscr{U}$ and put

$$\Theta(x, y) = \int_{\overline{x_0 x}} p_1(u, y)du^1 + \cdots + p_n(u, y)du^n, \tag{2.134}$$

where $u = x_0 + t(x - x_0)$ and $u = (u^1, \cdots, u^n)$. It follows that $du^j = (x^j - x_0^j)dt,\ j = 1, \cdots, n$ where $x = (x^1, \cdots, x^n)$ and $x_0 = (x_0^1, \cdots, x_0^n)$. Together with (2.134), we have

$$\Theta(x, y) = \int_0^1 \Big[(x^1 - x_0^1)p_1(t(x - x_0) + x_0, y) + \cdots$$
$$+(x^n - x_0^n)p_n(t(x - x_0) + x_0, y)\Big]dt.$$

It follows that Θ is differentiable with respect to y. Moreover we have

$$\begin{aligned}
\frac{\partial \Theta}{\partial x^i} &= \frac{\partial}{\partial x^i}\int_0^1 \sum_{j=1}^n (x^j - x_0^j)p_j(t(x - x_0) + x_0, y)dt \\
&= \int_0^1 \frac{\partial}{\partial x^i}\left[\sum_{j=1}^n (x^j - x_0^j)p_j(t(x - x_0) + x_0, y)\right]dt \\
&= \int_0^1 \sum_{j=1}^n \frac{\partial}{\partial x^i}\Big[(x^j - x_0^j)p_j(t(x - x_0) + x_0, y)\Big]dt \\
&= \int_0^1 \sum_{j=1}^n \Big[\delta_i^j p_j(t(x - x_0) + x_0, y) + t(x^j - x_0^j)(p_j)_{u^i}(t(x - x_0) + x_0, y)\Big]dt \\
&= \int_0^1 \Big[p_i(t(x - x_0) + x_0, y) + t\sum_{j=1}^n (x^j - x_0^j)(p_j)_{u^i}(t(x - x_0) + x_0, y)\Big]dt \\
&= \int_0^1 \frac{d}{dt}\Big[tp_i(t(x - x_0) + x_0, y)\Big]dt = tp_i(t(x - x_0) + x_0, y)\Big|_0^1 = p_i(x, y)
\end{aligned}$$

where we have used (2.133). Then we complete the proof of the Lemma 2.3.3.

Theorem 2.3.2 *Let $\mathcal{U}$ be an open subset in $\mathbb{R}^n$. Suppose that $F : T\mathcal{U} \to \mathbb{R}$ is a function which is positively homogeneous of degree one. Then $F = F(x, y)$ is a solution of* (2.1) *if and only if*

$$F^2 = \Theta_{x^i} y^i \tag{2.135}$$

where $\Theta : T\mathcal{U} \to \mathbb{R}$ satisfies the Hamel's equations (2.71).

Proof First suppose that F is a solution of (2.1). According to Lemma 2.3.1, there exists a function Θ such that (2.131) holds. Contracting (2.131) with y^i gives

$$\Theta_{x^i} y^i = F F_{y^i} y^i = F^2 = 2L. \tag{2.136}$$

Differentiating (2.136) with respect y^j, we obtain

$$2L_{y^j} = (\Theta_{x^i} y^i)_{y^j} = \Theta_{x^i y^j} y^i + \Theta_{x^j}. \tag{2.137}$$

Together with (2.131) yields (2.71).

Conversely, suppose that (2.135) holds, where $\Theta = \Theta(x, y)$ satisfies (2.71). Differentiating (2.135) with respect to x^j, we have $L_{x^j} = \left(\frac{F^2}{2}\right)_{x^j} = \frac{1}{2}\Theta_{x^i x^j} y^i$. It follows that

$$L_{x^j y^k} = \frac{1}{2}(I) + \frac{1}{2}\Theta_{x^k x^j}. \tag{2.138}$$

where

$$(I) := \Theta_{x^i x^j y^k} y^i = \Theta_{x^i y^k x^j} y^i = (\Theta_{x^i y^k} y^i)_{x^j} = \Theta_{x^k x^j} \tag{2.139}$$

where we have used (2.71). Plugging (2.139) into (2.138) yields $L_{x^j y^k} = \Theta_{x^k x^j}$. Note that $\Theta_{x^j x^k} = \Theta_{x^k x^j}$. Hence we obtain (2.95). Combining this with Lemma 2.2.3 we obtain F is a solution of dually flat equations (2.1).

Theorem 2.3.2 tells us that there is a bijection between solutions Θ of projectively flat equations (i.e. Hamel equations) and solutions F of dually flat equations, which are positively homogeneous of degree one, given by (2.140) and (2.139).

2.3.2 *Solution from Projectively Flat Equations*

In this subsection, we give an approach to manufacture solutions of (2.1) from projectively flat Finsler metrics in the form

$$\Phi(x, y) = |y|\left[\epsilon + \phi\left(|x|, \frac{\langle x, y\rangle}{|y|}\right)\right].$$

where $\epsilon > 0$.

Recall that a Finsler metric $F = F(x, y)$ on an open subset $\mathcal{U} \subset \mathbb{R}^n$ is said to be *projectively flat* if all geodesics are straight in $\mathcal{U}$.

Proposition 2.3.1 *Let* $\Phi(x, y) := |y|\left[\epsilon + \phi(|x|, \frac{\langle x, y\rangle}{|y|})\right]$ *be a projectively flat Finsler metric on an open subset* $\mathcal{U} \subset \mathbb{R}^n$. *Then the following function on* $T\mathcal{U}$

$$F(x, y) = |y|\left[\psi\left(|x|, \frac{\langle x, y\rangle}{|y|}\right)\right]^{\frac{1}{2}}. \tag{2.140}$$

is a solution of (2.1), *where* ψ *is given in* (2.146).

Proof By Hamel Lemma (see (2.71)), Φ is projectively flat if and only if it satisfies $\Phi_{x^j y^i} y^j = \Phi_{x^i}$. Together with Theorem 2.3.2 we have

$$2L := F^2 = \Phi_{x^j} y^j. \tag{2.141}$$

satisfies (2.1).

Now let us compute $\Phi_0 := \Phi_{x^i} y^i$ and F. Denote Φ by $\Phi = \Phi(r,s)$ where

$$r = |x|, \qquad s = \frac{\langle x, y\rangle}{|y|}. \tag{2.142}$$

By straightforward computations one obtains

$$\frac{\partial r}{\partial x^i} = \frac{x^i}{r}, \qquad \frac{\partial s}{\partial x^i} = \frac{y^i}{|y|}. \tag{2.143}$$

It follows that

$$\begin{aligned}\Phi_{x^i} &= \frac{\partial}{\partial x^i}\left[|y|\phi\left(|x|, \frac{\langle x, y\rangle}{|y|}\right)\right]\\ &= |y|\left(\frac{\partial\phi}{\partial r}\frac{\partial r}{\partial x^i} + \frac{\partial\phi}{\partial s}\frac{\partial s}{\partial x^i}\right) = |y|\left(\frac{x^i}{r}\frac{\partial\phi}{\partial r} + \frac{y^i}{|y|}\frac{\partial\phi}{\partial s}\right).\end{aligned} \tag{2.144}$$

Contracting (2.144) with y^i yields

$$\Phi_{x^i} y^i = |y|^2 \psi\left(|x|, \frac{\langle x, y\rangle}{|y|}\right), \tag{2.145}$$

where we have used (2.141) and ψ is defined by

$$\psi(r,s) := \frac{\partial\phi}{\partial s} + \frac{s}{r}\frac{\partial\phi}{\partial r}. \tag{2.146}$$

From (2.141), (2.151) and (2.152), one obtains

$$F = \sqrt{2L} = \left[|y|^2\left(\frac{\partial\phi}{\partial s} + \frac{s}{r}\frac{\partial\phi}{\partial r}\right)\right]^{\frac{1}{2}} = |y|\left[\psi\left(|x|, \frac{\langle x, y\rangle}{|y|}\right)\right]^{\frac{1}{2}} \tag{2.147}$$

which completes the proof of Proposition 2.3.1.

Taking $\phi(r,s) = \kappa + r^{\mu} f(\frac{s}{r})$ in Proposition 2.3.1 where κ and μ are constants, we have

$$\frac{\partial\phi}{\partial s} = r^{\mu-1} f'(\lambda), \tag{2.148}$$

$$\frac{\partial \phi}{\partial r} = \mu r^{\mu-1} f(\lambda) - s r^{\mu-2} f'(\lambda), \tag{2.149}$$

where

$$\lambda = \frac{s}{r} = \frac{\langle x, y \rangle}{|x||y|}. \tag{2.150}$$

Plugging (2.148) and (2.149) into (2.147) we obtain the following formula for F

$$F = |y| \left(\frac{\partial \phi}{\partial s} + \lambda \frac{\partial \phi}{\partial r} \right)^{\frac{1}{2}} = |y| \left\{ |x|^{\mu-1} \left[\mu \lambda f(\lambda) + (1 - \lambda^2) f'(\lambda) \right] \right\}^{\frac{1}{2}}.$$

Hence we obtain the following:

Corollary 2.3.1 *Let* $\Phi(x, y) := |y| \left[\epsilon + |x|^{\mu} f(\frac{\langle x, y \rangle}{|x||y|}) \right]$ *be a projectively flat Finsler metric on an open subset* $\mathscr{U} \subset \mathbb{R}^n \backslash \{0\}$*. Then the following function on* $T\mathscr{U}$

$$F(x, y) := |y| \left\{ |x|^{\mu-1} \left[\mu \lambda f(\lambda) + (1 - \lambda^2) f'(\lambda) \right] \right\}^{\frac{1}{2}}$$

is a solution of (2.1) *where* $\lambda = \frac{\langle x, y \rangle}{|x||y|}$.

2.3.3 New Dually Flat Finsler Metrics

In this subsection we are going to produce new dually flat Finsler metrics from a given projectively flat Finsler metric.

Lemma 2.3.3 *Let* $\Phi(x, y) := |y| \left[\epsilon + |x|^{\mu} f(\frac{\langle x, y \rangle}{|x||y|}) \right]$ *be a projectively flat Finsler metric on an open subset* $\mathscr{U} \subset \mathbb{R}^n \backslash \{0\}$*. Suppose that* $f(-1) < 0$*. Then*

$$F(x, y) := |y| |x|^{\frac{\mu-1}{2}} \left[\mu \lambda f(\lambda) + (1 - \lambda^2) f'(\lambda) \right]^{\frac{1}{2}}$$

is dually flat Finsler metric where $\mu > 0$.

Proof In fact, F is expressed in the form $F = |y| \phi(r, s),\ r = |x|,\ s = \frac{\langle x, y \rangle}{|y|}$ where

$$\phi = r^{\frac{\mu-1}{2}} \sqrt{\mu \lambda f(\lambda) + (1 - \lambda^2) f'(\lambda)}, \tag{2.151}$$

and λ satisfies (2.150). Further, F satisfies (2.1) by Corollary 2.3.1. It is known that $F = |y| \phi(r, s)$ is a Finsler metric with $r < b_0$ if and only if ϕ is a positive function satisfying

$$\phi(s) - s\phi_s(s) > 0, \quad \phi(s) - s\phi_s(s) + (r^2 - s^2)\phi_{ss}(s) > 0, \quad |s| \le r < b_0$$

where $n \ge 3$ or

$$\phi(s) - s\phi_s(s) + (r^2 - s^2)\phi_{ss}(s) > 0, \quad |s| \le r < b_0$$

where $n = 2$ (see Sect. 1.3). Note that Φ is projectively flat. From [27], we have

$$(\lambda^2 - 1)f'' - \mu\lambda f' + \mu f = 0 \tag{2.152}$$

and

$$f''(\lambda) = \mu(1 - \lambda^2)^{\frac{\mu}{2}-1}. \tag{2.153}$$

Differentiating (2.151) with respect to s, we obtain

$$\begin{aligned} 2\phi\phi_s &= r^{\mu-1}\frac{\partial\lambda}{\partial s}\left[\mu f + \mu\lambda f' - 2\lambda f' + (1+\lambda^2)f''\right] \\ &= r^{\mu-2}\left[\mu f + \mu\lambda f' - 2\lambda f' + \mu(f - \lambda f')\right] = 2r^{\mu-2}(\mu f - \lambda f') \end{aligned}$$

where we have used (2.150) and (2.152). It follows that

$$\phi_s = \frac{r^{\mu-2}(\mu f - \lambda f')}{\phi}. \tag{2.154}$$

Together with (2.151) and (2.154), we have

$$\phi - s\phi_s = \frac{1}{\phi}\left[\phi^2 - sr^{\mu-2}(\mu f - \lambda f')\right] = \frac{r^{\mu-1}}{\phi}f'. \tag{2.155}$$

Differentiating (2.154) with respect to s and using (2.154) one deduces

$$\phi_{ss} = \frac{(\mu-1)r^{\mu-3}}{\phi}f' - \frac{\lambda r^{\mu-3}}{\phi}f'' - \frac{r^{2\mu-4}}{\phi^3}(\mu f - \lambda f')^2.$$

Together with (2.150) and (2.152), we obtain

$$(r^2 - s^2)\phi_{ss} = \frac{r^{\mu-1}}{\phi}\left[\mu - (1-\lambda^2)\right]f' - \lambda\mu\frac{r^{\mu-1}}{\phi}f - \frac{r^{2\mu-2}}{\phi^3}(1-\lambda^2)(\mu f - \lambda f')^2.$$

Combining this with (2.155), we get

$$
\begin{aligned}
&\phi(s) - s\phi_s(s) + (r^2 - s^2)\phi_{ss}(s) \\
&= \frac{r^{\mu-1}}{\phi}\Big[(\mu + \lambda^2)f' - \lambda\mu f\Big] - \frac{r^{2\mu-2}}{\phi^3}(1-\lambda^2)(\mu f - \lambda f')^2 \\
&= \frac{r^{2\mu-2}}{\phi^3} \times (I)
\end{aligned} \tag{2.156}
$$

where

$$
\begin{aligned}
(I) &= \Big[\mu\lambda f + (1-\lambda^2)f'\Big]\Big[(\mu+\lambda^2)f' - \lambda\mu f\Big] \\
&\quad -(1-\lambda^2)\left(\mu^2 f^2 - 2\mu\lambda f f' + \lambda^2 f'^2\right) \\
&= \mu\Big[(1-\lambda^2)f'f' + (1+\mu)\lambda f f' - \mu f^2\Big].
\end{aligned} \tag{2.157}
$$

Plugging (2.157) into (2.156) yields

$$
\begin{aligned}
&\phi(s) - s\phi_s(s) + (r^2 - s^2)\phi_{ss}(s) \\
&= \frac{\mu r^{2\mu-2}}{\phi^3}\Big[(1-\lambda^2)f'f' + \lambda(1+\mu)ff' - \mu f^2\Big].
\end{aligned} \tag{2.158}
$$

By (2.151), (2.155) and (2.158), $F = |y|\phi(r, s)$ is a Finsler metric if and only if

$$
f' > 0, \tag{2.159}
$$

$$
g := \mu\lambda f + (1-\lambda^2)f' > 0, \tag{2.160}
$$

$$
h := (1-\lambda^2)f'f' + \lambda(1+\mu)ff' - \mu f^2 > 0 \tag{2.161}
$$

where $n \geq 3$ or, (2.160) and (2.161) hold when $n = 2$. By using (2.150) and Cauchy-Buniakowski inequality we are going to find conditions on f for (2.159), (2.160) and (2.161) to hold in $[-1, 1]$.

Note that $\mu > 0$. Together with (2.153) we get

$$
f''(\lambda) > 0 \tag{2.162}
$$

where $\lambda \in (-1, 1)$. It follows that $f'(\lambda)$ is a monotonically increasing function on $[-1, 1]$. Thus

$$
f'(-1) > 0 \tag{2.163}
$$

implies that (2.159) holds in $[-1, 1]$. Plugging (2.153) into (2.152) yields

$$
\mu(f - \lambda f') = (1-\lambda^2)f'' = (1-\lambda^2)\mu(1-\lambda^2)^{\frac{\mu}{2}-1} = \mu(1-\lambda^2)^{\frac{\mu}{2}}.
$$

It follows that $f - \lambda f' = (1-\lambda^2)^{\frac{\mu}{2}}$, which immediately implies that

$$f(1) = f'(1), \quad f(-1) = -f'(-1). \tag{2.164}$$

This means that (2.159) holds in $[-1, 1]$ if $f(-1) < 0$.

Next, we are going to a find condition on f for (2.160) to hold in $[-1, 1]$. By using (2.152) and (2.160), we have $g'(\lambda) = 2\left[\mu f(\lambda) - \lambda f'(\lambda)\right]$. It follows that $g'(\lambda) = 0$ if and only if

$$\mu f(\lambda) = \lambda f'(\lambda). \tag{2.165}$$

Suppose that $\lambda_0 \in [-1, 1]$ such that $g'(\lambda_0) = 0$. Combining this with (2.165), we have

$$\mu f(\lambda_0) = \lambda_0 f'(\lambda_0). \tag{2.166}$$

Together with (2.160) we get

$$g(\lambda_0) = \mu\lambda_0 f(\lambda_0) + (1-\lambda_0^2) f'(\lambda_0) = f'(\lambda_0). \tag{2.167}$$

On the other hand, from (2.160) and (2.164), we obtain

$$g(1) = \mu f'(1), \quad g(-1) = \mu f'(-1). \tag{2.168}$$

It is known that the minimum of g satisfies the following

$$\min_{\lambda\in[-1,1]} g(\lambda) = \min\left\{g(\lambda_0),\ g(\pm 1) \mid g'(\lambda_0) = 0\right\}.$$

It is easy to see that (2.160) holds for $\lambda \in [-1, 1]$ if and only if

$$\min_{\lambda\in[-1,1]} g(\lambda) > 0. \tag{2.169}$$

By (2.167) and (2.168), (2.169) holds if and only if

$$\min\left\{\mu f'(-1), f'(\lambda_0),\ \mu f'(1)\right\} > 0. \tag{2.170}$$

where $\lambda_0 \in [-1, 1]$ satisfying $g'(\lambda_0) = 0$. Note that $\mu > 0$ and f' is a monotonically increasing function. Together with the second equation of (2.164), we obtain that (2.170) holds if and only if $f(-1) < 0$.

Finally, we are going to find a condition on f for (2.161) to hold in $[-1, 1]$. Using (2.158) and (2.161) we get

$$h'(\lambda) = (1-\mu) f'(f - \lambda f') + (1+\mu)\lambda f f'' + 2(1-\lambda^2) f' f''$$

$$= (1-\mu)f'\frac{(1-\lambda^2)f''}{\mu} + (1+\mu)\lambda f f'' + 2(1-\lambda^2)f'f''$$

$$= \frac{\mu+1}{\mu} f''\left[\mu\lambda f + (1-\lambda^2)f'\right]. \tag{2.171}$$

By (2.161) and the second equation of (2.164), we see that

$$h(-1) = -(1+\mu)f(-1)f'(-1) - \mu\,[f(-1)]^2 = [f(-1)]^2. \tag{2.172}$$

Suppose that $f(-1) < 0$. Together with (2.172) yields

$$h(-1) > 0. \tag{2.173}$$

Moreover, (2.160) holds where $\lambda \in [-1, 1]$. Combining this with (2.162) and (2.171), we have $h'(\lambda) > 0, \quad \lambda \in (-1, 1)$. It follows that $h(\lambda)$ is a monotonically increasing function. Together with (2.173) ones obtain that (2.161) is true.

In [27], authors gave an explicit construction of projectively flat spherically symmetric Finsler metric (see Proposition 5.1 below). Precisely, they have proved the following:

Proposition 2.3.2 *Let $f(\lambda)$ be a polynomial function defined by*

$$f(\lambda) = 1 + \delta\lambda + 2n\sum_{k=0}^{n-1}\frac{(-1)^k C_{n-1}^k \lambda^{2k+2}}{(2k+1)(2k+2)}. \tag{2.174}$$

Then the following Finsler metric on open subset in $\mathbb{R}^n\backslash\{0\}$

$$\Phi = |y|\left\{\epsilon + |x|^{2n} f\left(\frac{\langle x, y\rangle}{|x||y|}\right)\right\}$$

is projectively flat where $\epsilon > 0$.

Proof of Theorem 2.3.1 Combine Lemma 2.3.3 with Proposition 2.3.2.

Chapter 3
Spherically Symmetric Metrics of Isotropic Berwald Curvature

For a Finsler metric $F = F(x, y)$ with spray coefficients $G^i = G^i(x, y)$, it is natural to consider the quantity given in (3.6). Because this quantity was introduced by Berwald first, we call it the *Berwald curvature* [65]. F is said to be of *isotropic Berwald metric* if its Berwald curvature $B_j{}^i{}_{kl}$ satisfies (3.7) for some scalar function $\sigma = \sigma(x)$ on the manifold M [21, 77]. Berwald metrics are trivially isotropic Berwald metric with $\sigma(x) = 0$. In this chapter, we are going to show that every spherically symmetric metric of isotropic Berwald curvature is a Randers metric. Then we shall also construct explicitly a lot of new isotropic Berwald spherically symmetric Finsler metrics.

3.1 Preliminaries

Let F be a Finsler metric on a manifold and $G^i = \frac{1}{4} g_{il} \left(\frac{\partial g_{jl}}{\partial x^k} + \frac{\partial g_{lk}}{\partial x^j} - \frac{\partial g_{jk}}{\partial x^l} \right) y^k y^j$ where g_{ij} are components of the fundamental tensor of F. We call G^i the *geodesic coefficients* of F.

Let $|\cdot|$ and $\langle\, ,\, \rangle$ be the standard Euclidean norm and inner product on $\mathbb{R}^n$. By a straightforward computation, one obtains the following lemma [82].

Lemma 3.1 *Let $F = |y|\phi(|x|, \frac{\langle x, y\rangle}{|y|})$ be a spherically symmetric Finsler metric on $\mathbb{B}^n(r_\mu) \subset \mathbb{R}^n$. Then its geodesic coefficients are given by*

$$G^i = uPy^i + u^2 Qx^i, \tag{3.1}$$

where

$$Q := \frac{1}{2r} \frac{r\phi_{ss} - \phi_r + s\phi_{rs}}{\phi - s\phi_s + (r^2 - s^2)\phi_{ss}}, \quad u := |y|, \quad r := |x|, \quad s := \frac{\langle x, y\rangle}{|y|} \tag{3.2}$$

E. Guo, X. Mo, *The Geometry of Spherically Symmetric Finsler Manifolds*, SpringerBriefs in Mathematics, https://doi.org/10.1007/978-981-13-1598-5_3

and

$$P := \frac{r\phi_s + s\phi_r}{2r\phi} - \frac{Q}{\phi}[s\phi - (r^2 - s^2)\phi_s]. \tag{3.3}$$

Let $F = F(x, y)$ be a Finsler metric on a manifold M of dimension n. Take an arbitrary standard local coordinate system (x^i, y^i) and define

$$\sigma_F(x) := \frac{Vol(B^n(1))}{Vol\{(y^i) \in \mathbb{R}^n | F(x, y^i(\partial/\partial x^i)|_x) < 1\}}.$$

For a non-zero vector $y \in T_x M$, the *distortion* $\tau = \tau(x, y)$ is defined by

$$\tau(x, y) := \ln \frac{\sqrt{det(g_{ij}(x, y))}}{\sigma_F(x)}.$$

Let $\gamma(t)$ be the geodesic with $\gamma(0) = x$ and $\dot{\gamma}(0) = y$. Let

$$\mathbf{S}(x, y) = \frac{d}{dt}[\tau(\gamma(t), \dot{\gamma}(t))]_{t=0},$$

where $\tau(x, y)$ is the distortion of F. $\mathbf{S}(x, y)$ is called the *S-curvature* [12, 13, 49]. F is said to have *isotropic S-curvature* if there is scalar function $c(x)$ on M such that

$$\mathbf{S} = (n + 1)c(x)F. \tag{3.4}$$

Lemma 3.2 ([10]) *Let $F := \alpha + \beta$ be the Randers metric on an n-dimensional manifold M where $\alpha = \sqrt{a_{ij}(x)y^i y^j}$ and $\beta = b_i(x)y^i$. Suppose that β is closed. Then F has isotropic S-curvature, i.e.* (3.4) *holds, if and only if*

$$b_{i|j} = 2c(x)(a_{ij} - b_i b_j). \tag{3.5}$$

3.2 Berwald Curvature of Spherically Symmetric Metrics

There is a set of local functions $B_j{}^i{}_{kl}$ on M defined by

$$B_j{}^i{}_{kl} := \frac{\partial^3 G^i}{\partial y^j \partial y^k \partial y^l}. \tag{3.6}$$

Because this quantity was introduced by L. Berwald first, we call it the *Berwald curvature* [65].

Definition 3.1 A Finsler metric F on M is said to be of *isotropic Berwald curvature* if its Berwald curvature $B_j{}^i{}_{kl}$ satisfies the following

$$B_j{}^i{}_{kl} = \sigma(x)(F_{y^j y^k}\delta^k_l + F_{y^k y^l}\delta^i_j + F_{y^l y^j}\delta^i_k + F_{y^j y^k y^l}y^i), \tag{3.7}$$

for some scalar function $\sigma = \sigma(x)$ on M[77].

Let $M = Bn(r_\mu)$. From (3.1) and (3.2), we have

$$\frac{\partial G^i}{\partial y^j} = u_{y^j}Py^i + uP_s s_{y^j}y^i + uP\delta^i_j + 2y^j Qx^i + u^2 Q_s s_{y^j}x^i, \tag{3.8}$$

where we have used $\frac{\partial r}{\partial y^i} = 0$ and $\frac{\partial u^2}{\partial y^j} = 2y^j$. By (3.8), we obtain

$$\begin{aligned}\frac{\partial^2 G^i}{\partial y^j \partial y^k} =&(P_s s_{y^k}y^i u_{y^j} + P\delta^i_k u_{y^j} + P_s u\delta^i_k s_{y^j} + 2Q_s y^k x^i s_{y^j})(j \leftrightarrow k)\\ &+ P_{ss}uy^i s_{y^j y^k} + P_s uy^i s_{y^j y^k} + Py^i u_{y^j y^k} + Q_{ss}u^2 x^i s_{y^j}s_{y^k}\\ &+ Q_s u^2 x^i s_{y^j y^k} + 2Qx^i\delta_{jk},\end{aligned}$$

where $j \leftrightarrow k$ denotes symmetrization, from which together with (3.6) we obtain

$$\begin{aligned}B_j{}^i{}_{kl} =&(P_{ss}y^i s_{y^j}s_{y^k}u_{y^l} + P_s y^i s_{y^j y^k}u_{y^l} + P_s y^i s_{y^j}u_{y^k y^l})(j \to k \to l \to j)\\ &+ [P_s(s_{y^j}u_{y^k} + s_{y^k}u_{y^j})\delta^i_l + Pu_{y^j y^k} + P_{ss}us_{y^j}s_{y^k}\delta^i_l](j \to k \to l \to j)\\ &+ (P_s us_{y^j y^k}\delta^i_l + P_{ss}uy^i s_{y^j}s_{y^k y^l} + Q_{ss}u^2 x^i s_{y^j}s_{y^k y^l})(j \to k \to l \to j)\\ &+ 2x^i(Q_s s_{y^j}\delta_{kl} + Q_{ss}y^j s_{y^k}s_{y^l} + Q_s y^j s_{y^k y^l})(j \to k \to l \to j)\\ &+ P_{sss}y^i us_{y^j}s_{y^k}s_{y^l} + Py^i u_{y^j y^k y^l} + P_s uy^i s_{y^i y^j y^k}\\ &+ Q_{sss}u^2 x^i s_{y^j}s_{y^k}s_{y^l} + Q_s u^2 x^i s_{y^j y^k y^l},\end{aligned} \tag{3.9}$$

where $j \to k \to l \to j$ denotes cyclic permutation.

Observe that

$$u_{y^j} = \frac{y^j}{u}, \tag{3.10}$$

$$u_{y^j y^k} = \frac{u^2\delta_{jk} - y^j y^k}{u^3}, \tag{3.11}$$

$$u_{y^j y^k y^l} = \frac{3y^j y^k y^l - u^2\delta_{jk}y^l(j \to k \to l \to j)}{u^5}, \tag{3.12}$$

where we have used (3.2). Direct computations yield

$$s_{y^j} = \frac{ux^j - sy^j}{u^2}, \tag{3.13}$$

$$s_{y^j y^k} = \frac{3sy^j y^k - ux^j y^k - ux^k y^j - su^2\delta_{jk}}{u^4}, \tag{3.14}$$

$$s_{y^j y^k y^l} = \frac{1}{u^5}(3x^j y^k y^l + 3su\delta_{jk} y^l - u^2 x^j \delta_{kl})(j \to k \to l \to j) - \frac{15}{u^6} s y^j y^k y^l. \tag{3.15}$$

From (3.9), (3.10), (3.11), (3.12), (3.13), (3.14) and (3.15) we get the following proposition.

Proposition 3.1 ([56]) *Let $F = |y|\phi(|x|, \frac{\langle x, y\rangle}{|y|})$ be a spherically symmetric Finsler metric on $\mathbb{B}^n(r_\mu) \subseteq \mathbb{R}^n$. Then the Berwald curvature of F is given by*

$$\begin{aligned}
B_j{}^i{}_{kl} =& \frac{1}{u}[P_{ss}\delta^i_j x^k x^l + (P - sP_s)\delta^i_j \delta_{kl}](j \to k \to l \to j) \\
&- \frac{sP_{ss}}{u^2}[\delta^i_j(x^k y^l + x^l y^k) + y^i \delta_{jk} x^l](j \to k \to l \to j) \\
&+ \frac{1}{u^3}(sP_{ss} + sP_s - P)(\delta^i_j y^k y^l + \delta_{jk} y^i y^l)(j \to k \to l \to j) \\
&+ \frac{1}{u^5}(3P - s^2 P_{sss} - 6s^2 P_{ss} - 3sP_s) y^i y^j y^k y^l + \frac{P_{sss}}{u^2} y^i x^j x^k x^l \\
&+ \frac{s}{u^4}(sP_{sss} + 3P_{ss}) y^i y^j y^k x^l (j \to k \to l \to j) \\
&+ \frac{1}{u}[(Q_s - sQ_{ss}) x^i x^l \delta_{jk} - \frac{1}{u^2}(P_{ss} + sP_{sss}) y^i y^j x^k x^l](j \to k \to l \to j) \\
&+ \frac{1}{u^3}(s^2 Q_{sss} + sQ_{ss} - Q_s) x^i x^j y^k y^l (j \to k \to l \to j) \\
&+ \frac{s}{u^2}[(sQ_{ss} - Q_s) x^i y^j \delta_{kl} - Q_{sss} x^i x^j x^l y^k](j \to k \to l \to j) \\
&+ \frac{x^i}{u^4}[u^3 Q_{sss} x^j x^k x^l + s(3Q_s - 3sQ_{ss} - s^2 Q_{sss}) y^j y^k y^l],
\end{aligned} \tag{3.16}$$

where P and Q are defined in (3.3) and (3.2) respectively, and $j \to k \to l \to j$ denotes cyclic permutation.

3.3 Spherically Symmetric Metrics of Isotropic Berwald Curvature

In this section we are going to discuss necessary and sufficient condition for a spherically symmetric metric to be isotropic Berwald. Then we show that every spherically symmetric Finsler metric of isotropic Berwald curvature is of Randers type.

Proposition 3.2 *Let $F = |y|\phi(|x|, \frac{\langle x,y\rangle}{|y|})$ be a spherically symmetric Finsler metric on $\mathbb{B}^n(r_\mu) \subset \mathbb{R}^n$. Then F is isotropic Berwald, i.e.* (3.7) *holds for some scalar function $\sigma = \sigma(x)$ on $\mathbb{B}_n(r_\mu)$ if and only if*

$$P - sP_s = \sigma(x)(\phi - s\phi_s), \tag{3.17}$$

$$Q_s - sQ_{ss} = 0, \tag{3.18}$$

where Q and P are defined in (3.2) *and* (3.3) *respectively. In particular, F is Berwald if and only if $P - sP_s = Q_s - sQ_{ss} = 0$.*

Proof F can be rewritten as $F = u\phi(r, s)$, where $u = |y|$, $r = |x|$, $s = \frac{\langle x,y\rangle}{|y|}$. It follows that $F_{y^j} = u_{y^j}\phi + u\phi_s s_{y^j}$ and

$$F_{y^jy^k} = u_{y^jy^k}\phi + (u_{y^j}s_{y^k} + u_{y^k}s_{y^j})\phi_s + us_{y^j}s_{y^k}\phi_{ss} + us_{y^jy^k}\phi_s. \tag{3.19}$$

Plugging (3.10), (3.11), (3.13) and (3.14) into (3.19) yields

$$\begin{aligned} u^3F_{y^jy^k} =&(u\delta_{jk} - y^jy^k)\phi + [y^j(ux^k - sy^k) + y^k(ux^j - sy^j)]\phi_s \\ &+ (ux^j - sy^j)(ux^k - sy^k)\phi_{ss} \\ &+ [3sy^jy^k - u(x^jy^k + x^ky^j) - su^2\delta_{jk}]\phi_s \\ =&u^2(\phi - s\phi_s)\delta_{jk} + u^2\phi_{ss}x^jx^k - u\phi_{ss}(x^jy^k + x^ky^j) \\ &- (\phi - s\phi_s - s^2\phi_{ss})y^jy^k. \end{aligned} \tag{3.20}$$

First, suppose that $F = |y|\phi\left(|x|, \frac{\langle x,y\rangle}{|y|}\right)$ is an isotropic Berwald metric. Then (3.7) holds. We take $j = k = l = i$, then $B_i{}^i{}_{ii} = \sigma(x)(3F_{y^iy^i} + F_{y^iy^iy^i}y^i)$, from which together with (3.16) and (3.20) we obtain

$$\begin{aligned} &\frac{3}{u}P_{ss}(x^i)^2 - \frac{3}{u}(sP_s - P) + \frac{3}{u}(Q_s - sQ_{ss})(x^i)^2 \\ &\equiv \sigma(x)\left[\frac{3}{u}(\phi - s\phi_s) + \frac{3}{u}\phi_{ss}(x^i)^2\right], \text{mod } y^i. \end{aligned} \tag{3.21}$$

This gives

$$P - sP_s = \sigma(x)(\phi - s\phi_s), \tag{3.22}$$

$$P_{ss} + Q_s - sQ_{ss} = \sigma(x)\phi_{ss}. \tag{3.23}$$

Moreover, σ is a function of $r := |x|$. It follows that

$$\frac{\partial \sigma}{\partial s} = 0. \tag{3.24}$$

Taking x and y with $x \wedge y \neq 0$ (cf.[27]). Differentiating (3.22) with respect to s and using (3.24), one obtains

$$P_{ss} = \sigma(x)\phi_{ss}. \tag{3.25}$$

Substituting (3.25) into (3.23) yields (3.18).

Conversely, suppose that F satisfies (3.17) and (3.18), then (3.24) holds. Combining this with (3.17), we have $\psi = s\psi_s$ where $\psi(r, s) = P - \sigma(x)\phi$. It is easy to see that the solution of ψ is $c(r)s$, i.e.

$$P - \sigma(x)\phi = c(r)s. \tag{3.26}$$

By (3.18), $Q = Q(r, s)$ is given by

$$Q = \frac{1}{2}a(r)s^2 + b(r). \tag{3.27}$$

Using (3.1), (3.26) and (3.27), we obtain

$$\begin{aligned} G^i - \sigma(x)Fy^i &= uPy^i + u^2Qx^i - \sigma Fy^i \\ &= u[P - \sigma(x)\phi]y^i + u^2[\frac{1}{2}a(r)s^2 + b(r)]x^i \\ &= uc(r)sy^i + \frac{1}{2}(us)^2 a(r)x^i + b(r)x^i|y|^2 \\ &= c(r)\langle x, y\rangle y^i + \frac{1}{2}a(r)\langle x, y\rangle^2 x^i + b(r)x^i|y|^2. \end{aligned} \tag{3.28}$$

Therefore $G^i - \sigma(x)Fy^i$ are quadratic in $y = y^j \frac{\partial}{\partial x^j}|_x$. On the other hand, using (3.6) and (3.24), we have

$$[G^i - \sigma(x)Fy^i]_{y^j y^k y^l} = B_j{}^i{}_{kl} - \sigma(x)(F_{y^j y^k}\delta^i_l + F_{y^k y^l}\delta^i_j + F_{y^l y^j}\delta^i_k + F_{y^j y^k y^l}y^i).$$

Combining this with (3.28), we get $F = |y|\phi\left(|x|, \frac{\langle x, y\rangle}{|y|}\right)$ has isotropic Berwald curvature. This completes the proof of Proposition 3.2.

Let $F = u\phi(r,s)$ be a spherically symmetric Finsler metric with isotropic Berwald curvature. From Proposition 3.2 and its proof, we see that (3.26) and (3.27) hold. Furthermore, $G^i - \sigma(x)Fy^i$ are quadratic in y, for some scalar function $\sigma : \mathbb{B}^n(r_\mu) \to \mathbb{R}$. Plugging (3.26) and (3.27) into (3.3) yields

$$\sigma(x)\phi + c(r)s = \frac{r\phi_s + s\phi_r}{2r\phi} - \frac{1}{\phi}[s\phi + (r^2 - s^2)\phi_s]\left[\frac{1}{2}a(r)s^2 + b(r)\right]. \tag{3.29}$$

Denote $a(r)$, $b(r)$ and $c(r)$ by a, b and c, respectively. Then (3.29) simplifies to

$$[(r^2-s^2)(2b+as^2)-1]r\phi_s - s\phi_r + rs(2b+as^2) + 2rcs\phi + 2r\sigma\phi^2 = 0. \tag{3.30}$$

Differentiating (3.30) with respect to s, we obtain

$$\begin{aligned}&[(r^2-s^2)(2b+as^2)-1]r\phi_{ss} - s\phi_{rs} - \phi_r + (3as^2+2c+2b)r\phi + 4r\sigma\phi\phi_s\\ &+ [2as(r^2-s^2) - s(2b+as^2) + 2cs]r\phi_s = 0.\end{aligned} \tag{3.31}$$

By using (3.2) and (3.27), we have $\frac{1}{2}as^2 + b = \frac{1}{2r}\frac{-\phi_r + s\phi_{rs} + r\phi_{ss}}{\phi - s\phi_s + (r^2-s^2)\phi_{ss}}$. It follows that

$$\left[(r^2-s^2)(2b+as^2)-1\right]r\phi_{ss} - s\phi_{rs} + \phi_r + r(2b+as^2)(\phi - s\phi_s) = 0. \tag{3.32}$$

By (3.31)–(3.32), we get

$$\begin{aligned}0 =& -2\phi_r + (3as^2+2c+2b)r\phi + 4r\sigma\phi\phi_s\\ &+\left[2a(r^2-s^2)s - s(2b+as^2) + 2cs\right]r\phi_s - r(2b+as^2)(\phi - s\phi_s)\\ =&[2a(r^2-s^2)s + 2cs]r\phi_s - 2\phi_r + (2as^2+2c)r\phi + 4r\sigma\phi\phi_s.\end{aligned}$$

Thus

$$[a(r^2-s^2)s + cs]r\phi_s - \phi_r + (as^2+c)r\phi + 2r\sigma\phi\phi_s = 0. \tag{3.33}$$

By (3.33)$\times s-$(3.30), we have $(2bs+cs+2\sigma\phi)\phi = [cs^2 - 2(r^2-s^2)b + 2s\sigma\phi + 1]\phi_s$. It follows that

$$(As^2 + 2\sigma s\phi + B)\phi_s = As\phi + 2\sigma\phi^2 \tag{3.34}$$

where

$$A := c + 2b = A(r), \quad B := 1 - 2br^2 = B(r). \tag{3.35}$$

For a fixed r, (3.34) is equivalent to the following equation

$$Nds + Md\phi = 0, \tag{3.36}$$

where $N := -(As\phi + 2\sigma\phi^2)$ and $M := As^2 + 2\sigma s\phi + B$. By simple calculations, we have $\frac{\partial M}{\partial s} = 2As + 2\sigma\phi$, $\frac{\partial N}{\partial \phi} = -As - 4\sigma\phi$. Hence

$$\frac{1}{N}\left(\frac{\partial M}{\partial s} - \frac{\partial N}{\partial \phi}\right) = \frac{3As + 6\sigma\phi}{-(As\phi + 2\sigma\phi^2)} = -\frac{3}{\phi}.$$

It follows that (3.36) has integrating factor which only depends on ϕ (see Theorem 2.5 in [19]). Moreover, its integrating factor is $\mu(\phi) = e^{-\int \frac{3}{\phi}d\phi} = \frac{1}{\phi^3}$. By (3.36)$\times\mu(\phi)$, we get the following exact equation

$$\frac{1}{\phi^3}\Big[(As\phi + 2\sigma\phi^2)ds - (As^2 + 2\sigma s\phi + B)d\phi\Big] = 0.$$

Thus we have the following general integral $A\frac{s^2}{\phi^2} + 4\sigma\frac{s}{\phi} + \frac{B}{\phi^2} = \lambda = \lambda(r)$. We obtain the quadratic equation

$$\lambda\phi^2 - 4\sigma s\phi - (As^2 + B) = 0. \tag{3.37}$$

Solving (3.37) for ϕ, we get $\phi = \frac{2\sigma s + \sqrt{(4\sigma^2 + \lambda A)s^2 + \lambda B}}{\lambda}$. Note that $s = \frac{\langle x, y\rangle}{|y|}$, $F = |y|\phi(r, s)$. It follows that

$$F = \frac{\sqrt{\lambda B|y|^2 + (4\sigma^2 + \lambda A)\langle x, y\rangle^2} + 2\sigma\langle x, y\rangle}{\lambda},$$

where A, B, σ and λ are functions of $|x|$. This means that F is a Randers metric. Therefore we have the following:

Theorem 3.1 *Let $(\mathbb{B}^n(r_\mu), F)$ be a spherically symmetric Finsler manifold. Suppose that F is of isotropic Berwald curvature. Then F is a Randers metric.*

3.4 Isotropic Berwald Metrics of Randers Type

Define α and β by

$$\alpha(x, y) = \sqrt{R(|x|)|y|^2 + S(|x|)\langle x, y\rangle^2}, \tag{3.38}$$

$$\beta(x, y) = T(|x|)\langle x, y\rangle. \tag{3.39}$$

It is easy to get the following:

Lemma 3.3 *Let $F := \alpha+\beta$ be any function given in* (3.38) *and* (3.39). *Then* 1*-form β is closed. Moreover, F is a Randers metric if and only if*

$$R(r) > \max\{0, r^2(T^2 - S)(r)\}, \tag{3.40}$$

where r is defined in (3.2).

Set

$$\alpha^2 = a_{ij}y^i y^j, \qquad \beta = b_i y^i. \tag{3.41}$$

Then

$$a_{ij} = R\delta_{ij} + Sx^i x^j, \qquad b_i = Tx^i. \tag{3.42}$$

We assume that $F = \alpha + \beta$ is of Randers type. Write $(a^{ij}) = (a_{ij})^{-1}$. By using (3.42) and Chern-Shen's Lemma 1.1.1 [17], we have

$$a^{ij} = \frac{1}{R}\left(\delta^{ij} - \frac{Sx^i x^j}{R + S|x|^2}\right). \tag{3.43}$$

From (3.2), we get

$$\frac{\partial r}{\partial x^k} = \frac{x^i}{r}. \tag{3.44}$$

Together with (3.42) we have

$$\frac{\partial a_{ij}}{\partial x^k} = \frac{1}{r}R_r x^k \delta_{ij} + \frac{1}{r}S_r x^i x^j x^k + S(\delta_{ik}x^j + \delta_{jk}x^i). \tag{3.45}$$

This implies that

$$\begin{aligned} \gamma_{ijk} :&= \frac{1}{2}\left(\frac{\partial a_{ij}}{\partial x^k} + \frac{\partial a_{ik}}{\partial x^j} - \frac{\partial a_{jk}}{\partial x^i}\right) \\ &= \tfrac{1}{2r}\left[R_r x^k \delta_{ij} + R_r x^j \delta_{ik} + (2rS - R_r)\delta_{jk}x^i + S_r x^i x^j x^k\right]. \end{aligned} \tag{3.46}$$

By using (3.43) and (3.46) we get

$$\begin{aligned} \gamma^i_{jk} &= a^{il}\gamma_{ljk} \\ &= \frac{1}{2rR}\left[R_r x^k \delta^i_j + R_r x^j \delta^i_k + (2rS - R_r)\delta_{jk}x^i + S_r x^i x^j x^k\right] \\ &\quad - \frac{x^i}{2rR(R + Sr^2)}\left[2R_r Sx^j x^k + (2rS - R_r)\delta_{jk}Sr^2 + SS_r x^i x^j x^k r^2\right]. \end{aligned} \tag{3.47}$$

Note that b_i satisfies the second equation of (3.42), we have

$$b_i \gamma^i_{jk} = \frac{T}{2(R+Sr^2)}\left[\left(\frac{2}{r}R_r + rS_r\right)x^j x^k + r(2rS - R_r)\delta_{jk}\right], \tag{3.48}$$

where we have used the following:

$$2R_r + S_r r^2 - 2\frac{SR_r r^2}{R+Sr^2} - \frac{SS_r r^4}{R+Sr^2} = R\frac{2R_r + S_r r^2}{R+Sr^2}$$

and

$$(2rS + R_r)r^2 - \frac{(2rS - R_r)Sr^4}{R+Sr^2} = R\frac{2rS - R_r}{R+Sr^2}r^2.$$

Furthermore,

$$\frac{\partial b_i}{\partial x^j} = \frac{1}{r}T_r x^i x^j + T\delta_{ij}. \tag{3.49}$$

Together with (3.48) we get

$$\begin{aligned} b_{i|j} :&= \frac{\partial b_i}{\partial x^j} - b_k \gamma^k_{ij} \\ &= \tfrac{1}{r}T_r x^i x^j + T\delta_{ij} - \frac{T}{2(R+Sr^2)}\left[\left(\frac{2}{r}R_r + rS_r\right)x^i x^j + r(2rS - R_r)\delta_{ij}\right] \\ &= \frac{T}{2}\frac{2R + rR_r}{R+Sr^2}\delta_{ij} + \frac{1}{2r}\left[2T_r - \frac{T(2R_r + r^2 S_r)}{R+Sr^2}\right]x^i x^j. \end{aligned} \tag{3.50}$$

From (3.42) we have

$$a_{ij} - b_i b_j = R\delta_{ij} + (S - T^2)x^i x^j. \tag{3.51}$$

From Lemma 3.3, β is closed. Together with Lemma 3.2, F has isotropic S-curvature if and only if (3.5) holds. By (3.50) and (3.51), (3.5) holds if and only if

$$\frac{T}{2}\frac{2R+rR_r}{R+Sr^2}\delta_{ij} + \frac{1}{2r}\left[2T_r - \frac{T(2R_r+r^2S_r)}{R+Sr^2}\right]x^i x^j = 2c(x)\left[R\delta_{ij} + (S - T^2)x^i x^j\right]. \tag{3.52}$$

Equation (3.52) holds if and only if functions R, S, and T satisfy

$$rT(S - T^2)(2R + r\frac{\partial R}{\partial r}) = R[2\frac{\partial T}{\partial r}(R + r^2 S) - T(2\frac{\partial R}{\partial r} + r^2\frac{\partial S}{\partial r})]. \tag{3.53}$$

In this case, we have

$$c(x) = c(|x|) = \frac{T}{4R}\frac{2R + rR_r}{R + Sr^2}.$$

Thus, we have the following:

Theorem 3.2 *Let $F = \alpha + \beta$ be the Randers metrics on $\mathbb{B}^n(r_\mu)$ defined by* (3.38) *and* (3.39). *Then F has isotropic S-curvature if and only if* (3.53) *holds. In this case, the S-curvature is given by*

$$\mathbf{S} = \frac{(n+1)T}{4R}\frac{2R + rR_r}{R + Sr^2}F. \tag{3.54}$$

Let us take a look at the special case: when $S = T^2$, the spherically symmetric Randers metric is given by

$$F = \sqrt{R(|x|)|y|^2 + [T(|x|)\langle x, y\rangle]^2} + T(|x|)\langle x, y\rangle.$$

By (3.53), F has isotropic S-curvature if and only if

$$0 = 2T_r(R + r^2T^2) - T\left[2R_r + r^2(T^2)_r\right] = 2(T_rR - TR_r). \tag{3.55}$$

By using Lemma 3.3, we have $R > 0$. It follows that (3.55) is equivalent to $\left(\frac{T}{R}\right)_r = 0$ i.e. $\frac{T}{R} = \kappa$ =constant. Combining this with (3.55) and rewriting $R(r)$ by $f(r)$ we obtain the following:

Theorem 3.3 *On $\mathbb{B}^n(r_\mu)$, the following spherically symmetric Randers metric*

$$F = \sqrt{f(|x|)|y|^2 + \kappa^2 f^2(|x|)\langle x, y\rangle^2} + \kappa f(|x|)\langle x, y\rangle$$

has isotropic Berwald curvature. Furthermore, its S-curvature is given by

$$\mathbf{S} = \frac{(n+1)\kappa}{4}\frac{2f(|x|) + |x|f_r(|x|)}{1 + \kappa^2|x|^2 f(|x|)}F,$$

where $r = |x|$, $f_r = \frac{\partial f}{\partial r}$, f is an any differentiable function and κ is a constant.

Finally, we show the following rigidity result.

Theorem 3.4 *Let $(\mathbb{B}^n(r_\mu), F)$ be a spherically symmetric Finsler manifold. Suppose that F is of isotropic Berwald curvature. Then F is the following Randers metric*

$$F = \sqrt{R(|x|)|y|^2 + S(|x|)\langle x, y\rangle^2} + T(|x|)\langle x, y\rangle, \tag{3.56}$$

where functions R, S and T satisfy (3.53) *and* $r = |x|$.

Proof Let F be a spherically symmetric Finsler metric of isotropic Berwald curvature on $\mathbb{B}^n(r_\mu)$. By Proposition 1.3.1, we have

$$F(x, y) = |y|\phi\left(|x|, \frac{\langle x, y\rangle}{|y|}\right),$$

where $\phi : [0, r_\mu) \times \mathbb{R} \to \mathbb{R}$. Together with Theorem 3.1, we obtain that F is given by (3.56). By Proposition 2.3 in [9], F has isotropic mean Berwald curvature, i.e. $E_{ij} = \frac{n+1}{2} F_{y^i y^j}$ where E is the *mean Berwald curvature* of F. It is easy to see that $E_{ij} := B_i{}^k{}_{jk}$. Combing this with Theorem 1.1 in [10], F is of isotropic S-curvature. By Theorem 3.3, R, S and T in (3.56) satisfy (3.53). Thus we complete the proof of the theorem.

Chapter 4
Spherically Symmetric Douglas Metrics

A Finsler metric F on an n-dimensional manifold M is a *Douglas metric* if its Douglas curvatures vanishes. The *Douglas curvature tensor* is defined by

$$D := D_j{}^i{}_{kl} dx^j \otimes \frac{\partial}{\partial x^i} \otimes dx^k \otimes dx^l,$$

where

$$D_j{}^i{}_{kl} := \frac{\partial^3}{\partial y^j \partial y^k \partial y^l} \left(G^i - \frac{1}{n+1} \sum_m \frac{\partial G^m}{\partial y^m} y^i \right) \tag{4.1}$$

in local coordinates $x^1, \cdots, x^n$ and $y = \sum_i y^i \frac{\partial}{\partial x^i}$ and $G^i = G^i(x, y)$ are geodesic coefficients of F. It was introduced by J. Douglas in 1927 [20]. Its importance in Finsler geometry is due to the fact that it is a projective invariant. Namely, if two Finsler metrics F and $\bar{F}$ are projectively equivalent, then F and $\bar{F}$ have the same Douglas curvature.

A Finsler metric on a manifold M is of Douglas type if its geodesic coefficients $G^i = G^i(x, y)$ are in the following form

$$G^i = \frac{1}{2} \Gamma^i_{jk}(x) y^j y^k + P(x, y) y^i$$

in local coordinates, where $\Gamma^i_{jk}(x)$ are functions on M, and $P(x, y)$ is a local positively y-homogeneous function of degree one.

In this chapter, we will obtain the differential equation that characterizes the spherically symmetric Finsler metrics with vanishing Douglas curvature. By solving this equation, we obtain all spherically symmetric Douglas metrics. Many explicit examples are included.

E. Guo, X. Mo, *The Geometry of Spherically Symmetric Finsler Manifolds*, SpringerBriefs in Mathematics, https://doi.org/10.1007/978-981-13-1598-5_4

4.1 Douglas Curvature of Spherically Symmetric Finsler Metrics

In our next result, we obtain the Douglas curvature of a spherically symmetric Finsler metric on $\mathbb{B}^n(r_\mu)$.

Proposition 4.1 *Let* $F = |y|\phi(|x|, \frac{\langle x, y\rangle}{|y|})$ *be a spherically symmetric Finsler metric on* $\mathbb{B}^n(r_\mu) \subset \mathbb{R}^n$. *Let* $u = |y|$ *and* $s = \frac{\langle x, y\rangle}{|y|}$, *then the Douglas curvature of* F *is given by*

$$\begin{aligned}
D_j{}^i{}_{kl} = & \frac{1}{u}[R_{ss}\delta^i_j x^k x^l + (R - sR_s)\delta^i_j \delta_{kl}](j \to k \to l \to j) \\
& -\frac{sR_{ss}}{u^2}[\delta^i_j(x^k y^l + x^l y^k) + y^i \delta_{jk} x^l](j \to k \to l \to j) \\
& +\frac{1}{u^3}(s^2 R_{ss} + sR_s - R)(\delta^i_j y^k y^l + y^i \delta_{jk} y^l)(j \to k \to l \to j) \\
& +\frac{1}{u^5}(3R - s^3 R_{sss} - 6s^2 R_{ss} - 3sR_s)y^i y^j y^k y^l + \frac{R_{sss}}{u^2} y^i x^j x^k x^l \\
& +\frac{s}{u^4}(sR_{sss} + 3R_{ss})y^i y^j y^k x^l (j \to k \to l \to j) \qquad (4.2) \\
& +\frac{1}{u}[(Q_s - sQ_{ss})x^i x^l \delta_{jk} - \frac{1}{u^2}(R_{ss} + sR_{sss})y^i y^j x^k x^l](j \to k \to l \to j) \\
& +\frac{1}{u^3}(s^2 Q_{sss} + sQ_{ss} - Q_s)x^i x^j y^k y^l (j \to k \to l \to j) \\
& +\frac{s}{u^2}[(sQ_{ss} - Q_s)x^i y^j \delta_{kl} - Q_{sss} x^i x^j x^l y^k](j \to k \to l \to j) \\
& +\frac{x^i}{u^4}[u^3 Q_{sss} x^j x^k x^l + s(3Q_s - 3sQ_{ss} - s^2 Q_{sss})y^j y^k y^l],
\end{aligned}$$

where Q *is given by* (3.2) *and*

$$R := -\frac{1}{n+1}\left[2sQ + (r^2 - s^2)Q_s\right]. \qquad (4.3)$$

Proof Let F be a spherically symmetric Finsler metric. From (3.1) and (3.2), we have

$$\begin{aligned}
\sum_{j=1}^{n} \frac{\partial G^j}{\partial y^j} &= \sum u_{y^j} P y^j + uP_s \sum s_{y^j} y^j + nuP + 2Q\langle x, y\rangle + u^2 Q_s \sum s_{y^j} x^j \\
&= u\left[(n+1)P + 2sQ + (r^2 - s^2)Q_s\right].
\end{aligned}$$

It follows that

$$G^i - \frac{1}{n+1}\sum_j \frac{\partial G^j}{\partial y^j} y^i = uRy^i + u^2 Qx^i,$$

where R is given by (4.3). Substituting $G^i - \frac{1}{n+1}\sum_j \frac{\partial G^j}{\partial y^j} y^i$ into (4.1) we get

$$D_j{}^i{}_{kl} = \frac{\partial^3}{\partial y^j \partial y^k \partial y^l}\left(uRy^i + u^2 Qx^i\right). \tag{4.4}$$

A straightforward computation implies that

$$\frac{\partial}{\partial y^j}\left(uRy^i + u^2 Qx^i\right) = u_{y^j} Ry^i + uR_s s_{y^j} y^i + uR\delta^i_j + 2y^j Qx^i + u^2 Q_s s_{y^j} x^i, \tag{4.5}$$

where we used $\frac{\partial r}{\partial y^i} = 0$ and $\frac{\partial u^2}{\partial y^j} = 2y^j$. From (4.5), we obtain

$$\begin{aligned}\frac{\partial^2}{\partial y^j \partial y^k}\left(uRy^i + u^2 Qx^i\right) = &(R_s s_{y^k} y^i u_{y^j} + R\delta^i_k u_{y^j} + R_s u\delta^i_k s_{y^j} + 2Q_s y^k x^i s_{y^j})(j \leftrightarrow k)\\ &+R_{ss} u y^i s_{y^j y^k} + R_s u y^i s_{y^j y^k} + R y^i u_{y^j y^k} + Q_{ss} u^2 x^i s_{y^j} s_{y^k}\\ &+Q_s u^2 x^i s_{y^j y^k} + 2Qx^i \delta_{jk},\end{aligned}$$

where $j \leftrightarrow k$ denotes symmetrization. Hence, it follows from (4.4) that

$$\begin{aligned}D_j{}^i{}_{kl} = &(R_{ss} y^i s_{y^j} s_{y^k} u_{y^l} + R_s y^i s_{y^j y^k} u_{y^l} + R_s y^i s_{y^j} u_{y^k y^l})(j \to k \to l \to j)\\ &+[R_s(s_{y^j} u_{y^k} + s_{y^k} u_{y^j})\delta^i_l + Ru_{y^j y^k}\delta^i_l + R_{ss} u s_{y^j} s_{y^k}\delta^i_l](j \to k \to l \to j)\\ &+(R_s u s_{y^j y^k}\delta^i_l + R_{ss} u y^i s_{y^j} s_{y^k y^l} + Q_{ss} u^2 x^i s_{y^j} s_{y^k y^l})(j \to k \to l \to j)\\ &+2x^i(Q_s s_{y^j}\delta_{kl} + Q_{ss} y^j s_{y^k} s_{y^l} + Q_s y^j s_{y^k y^l})(j \to k \to l \to j)\\ &+R_{sss} y^i u s_{y^j} s_{y^k} s_{y^l} + R y^i u_{y^j y^k y^l} + R_s u y^i s_{y^i y^j y^k}\\ &+Q_{sss} u^2 x^i s_{y^j} s_{y^k} s_{y^l} + Q_s u^2 x^i s_{y^j y^k y^l},\end{aligned} \tag{4.6}$$

where $j \to k \to l \to j$ denotes cyclic permutation. From (4.6), (3.10), (3.11), (3.12), (3.13), (3.14) and (3.15) we conclude the proof.

4.2 Spherically Symmetric Douglas Metrics

In this section we are going to discuss necessary and sufficient condition for a spherically symmetric metric to be a Douglas metric.

Lemma 4.1 *Let* $F = |y|\phi(|x|, \frac{\langle x, y\rangle}{|y|})$ *be a spherically symmetric Finsler metric on* $\mathbb{B}^n(r_\mu) \subset \mathbb{R}^n$. *Then* F *has vanishing Douglas curvature if and only if* (3.18) *holds.*

Proof Suppose that $F = |y|\phi\left(|x|, \frac{\langle x, y\rangle}{|y|}\right)$ is a Douglas metric, then $D_j{}^i{}_{kl} = 0$. We take $j = k = l = i$, then $D_i{}^i{}_{ii} = 0$, implies together with (4.6) that

$$\frac{3}{u}R_{ss}(x^i)^2 - \frac{3}{u}(sR_s - R) + \frac{3}{u}(Q_s - sQ_{ss})(x^i)^2 \equiv 0, \mod y^i.$$

This gives

$$R - sR_s = 0, \tag{4.7}$$

and

$$R_{ss} + Q_s - sQ_{ss} = 0. \tag{4.8}$$

Differentiating (4.7) with respect to s one obtains

$$R_{ss} = 0. \tag{4.9}$$

Substituting (4.9) into (4.8) we get (3.18).

Conversely, suppose that F satisfies (3.18) and (4.7). It is easy to see that

$$R = c(r)s. \tag{4.10}$$

From (3.18), $Q = Q(r, s)$ is given by

$$Q = \frac{1}{2}f(r)s^2 + g(r). \tag{4.11}$$

Using (4.10) and (4.11), we obtain

$$\begin{aligned} G^i - \tfrac{1}{n+1}\textstyle\sum_j \frac{\partial G^j}{\partial y^j} y^i &= uRy^i + u^2Qx^i \\ &= uc(r)sy^i + \tfrac{1}{2}(us)^2 f(r)x^i + g(r)x^i|y|^2 \\ &= c(r)\langle x, y\rangle y^i + \tfrac{1}{2}f(r)\langle x, y\rangle^2 x^i + g(r)x^i|y|^2. \end{aligned}$$

Hence, $G^i - \frac{1}{n+1}\sum_j \frac{\partial G^j}{\partial y^j} y^i$ is quadratic in $y = y^j \frac{\partial}{\partial x^j}|_x$. Combining this with (4.1), we get $F = |y|\phi\left(|x|, \frac{\langle x, y\rangle}{|y|}\right)$ has vanishing Douglas curvature.

Therefore, we have proved that (3.18) and (4.7) are necessary and sufficient conditions for the vanishing of the Douglas curvature of F.

Differentiating (4.3) with respect to s, one obtains $R_s = -\frac{1}{n+1}\left[2Q + (r^2 - s^2)Q_{ss}\right]$. It follows that $R - sR_s = -\frac{r^2-s^2}{n+1}(Q_s - sQ_{ss})$, i.e., (3.18) implies (4.7). Hence, we conclude that F has vanishing Douglas curvature if and only if, (3.18) holds.

Combining (4.11) with the first equation of (3.2), we get the following:

Theorem 4.1 *On $\mathbb{B}^n(r_\mu)$, a spherically symmetric Finsler metric $F(x, y) = |y|\phi(r, s)$ is of Douglas type if and only if, ϕ satisfies*

$$[(r^2 - s^2)(2g + fs^2) - 1]r\phi_{ss} - s\phi_{rs} + \phi_r + r(2g + fs^2)(\phi - s\phi_s) = 0, \quad (4.12)$$

where r and s are defined in (3.2), $f = f(r)$ and $g = g(r)$ are arbitrary differentiable functions.

One can show that under generic conditions, the differential equation (4.12) is equivalent to a transport equation. Transport equations arise in many mathematical problems and, in particular, in most PDEs related to fluid mechanics. By using the characteristic curves, we will provide the general solution of (4.12).

Let $f(r)$ and $g(r)$ be functions such that the following integrals are well defined for $r < r_\mu$

$$\int 2r(2g + r^2 f)dr \quad \text{and} \quad \int 2rfe^{\int 2r(2g+r^2 f)}dr. \quad (4.13)$$

In Sect. 4.3, we prove the following:

Theorem 4.2 *Let $f(r)$ and $g(r)$ be differential functions of $r \in I \subset \mathbb{R}$ such that conditions (4.13) hold. Then the general solution of (4.12), when $r^2 - s^2 > 0$ and $s \neq 0$ is*

$$\phi(r, s) = s(h(r) - \int \frac{\eta(\varphi(r, s))}{s^2\sqrt{r^2 - s^2}}ds), \quad (4.14)$$

where

$$\varphi(r, s) = \frac{r^2 - s^2}{(r^2 - s^2)\int 2rfe^{\int 2r(2g+r^2 f)dr}dr - e^{\int 2r(2g+r^2 f)dr}} \quad (4.15)$$

h and η are arbitrary differentiable real functions of r and φ respectively. Moreover, any spherically symmetric Douglas metric on $\mathbb{B}^n(r_\mu)$ is given by

$$F(x, y) = |y|\phi(|x|, \frac{\langle x, y\rangle}{|y|}),$$

where ϕ is of the form (4.14) and

$$\frac{-\sqrt{r^2 - s^2}}{s}\frac{\partial \eta}{\partial s} > 0, \quad when \quad n \geq 2, \quad (4.16)$$

with the additional inequality

$$\frac{\eta}{\sqrt{r^2 - s^2}} > 0, \quad when \quad n \geq 3. \quad (4.17)$$

4.3 General Solution

In this section, we will prove Theorem 4.2 providing the general solution of (4.12), under the generic conditions (4.13) for the functions $f(r)$ and $g(r)$. We will use the method of characteristic curves to obtain the general solution of the so called transport equation [60] with non-constant coefficients

$$\psi_r(r, s) + v(r, s)\psi_s(r, s) = P(r, s, \psi(r, s)). \tag{4.18}$$

As we will see in the proof of Theorem 4.2, Eq. (4.12) is equivalent to an equation of type (4.18). Observe that this equation can be written as $\mathbf{v} \cdot \nabla\psi = P$, where $\mathbf{v} = (1, v)$ and $\nabla\psi = (\psi_r, \psi_s)$. This equation thus has a geometric interpretation: we seek a surface $z = \psi(r, s)$ whose directional derivative in the direction of vector $\mathbf{v}$ is $P(r, s, \psi)$. This geometric interpretation is the basis for the following method for solving (4.18).

Curves $(r, X(r))$ in the (r, s)-plane that are tangential to the vector field $(1, v)$ are called *characteristic curves*.

It follows from this definition, that the characteristic curve that goes through the point $(r, s) = (r_0, c)$ is the graph of the function X that satisfies the ODE

$$\frac{dX}{dr} = v(r, X(r)), \tag{4.19}$$

with initial condition $X(r_0) = c$.

Denoting the value of ψ, along a characteristic curve $(r, X(r))$, by $\Psi(r) = \psi(r, X(r))$, we have

$$\frac{d}{dr}\Psi = \frac{\partial\psi}{\partial r} + \frac{\partial\psi}{\partial s}\frac{dX}{dr} = \psi_r + v\psi_s = P. \tag{4.20}$$

Hence, the value of ψ along a characteristic curve is determined by the ODE

$$\Psi' = P(r, X(r), \Psi(r)). \tag{4.21}$$

The solution of the ODE (4.21), with initial value $\Psi(r_0) = \psi_{r_0}(c)$ determines the value of ψ along the characteristic curve that intersects the (r_0, s)-axis at (r_0, c), because $\Psi(r_0) = \psi(r_0, X(0)) = \psi(r_0, c) = \psi_{r_0}(c)$. The surface $z = \psi(r, s)$ is the collection (or envelope) of space curves created as c takes on all real values.

Proof of Theorem 4.2: Since $r^2 - s^2 > 0$ and $s \neq 0$, equation (4.12) is equivalent to:

$$\left[1 - (r^2 - s^2)(2g(r) + f(r)s^2)\right] r\psi_s(r, s) + s\psi_r(r, s) = 0, \tag{4.22}$$

where

$$\psi = (\phi - s\phi_s)\sqrt{r^2 - s^2}. \tag{4.23}$$

Since $s \neq 0$, observe that (4.22) is equivalent to the following transport equation

$$\psi_r + v(r, s)\psi_s = 0, \tag{4.24}$$

$$\psi(r_0, s) = \psi_{r_0}(s), \tag{4.25}$$

where

$$v(r, s) = \frac{r}{s}\left[1 - (r^2 - s^2)(2g(r) + f(r)s^2)\right]. \tag{4.26}$$

We consider the characteristic curves $(r, X(r))$ of this equation. By defining $\kappa(r) = X^2(r) - r^2$, we rewrite equation (4.19) as

$$\kappa' = 2r(2g + r^2 f)\kappa + 2rf\kappa^2. \tag{4.27}$$

which is a Ricatti type equation whose solution is

$$\kappa = \frac{e^{\int_{r_0} 2r(2g+r^2 f)dr}}{c_0 - \int_{r_0} 2rfe^{\int_{r_0} 2r(2g+r^2 f)dr}dr}, \qquad c_0 \in \mathbb{R}. \tag{4.28}$$

Then, the solution of equation (4.19) with initial condition $X(r_0) = c$ is:

$$X(r) = \sqrt{r^2 + \frac{e^{\int_{r_0} 2r(2g+r^2 f)dr}}{\frac{1}{c^2 - r_0^2} - \int_{r_0} 2rfe^{\int_{r_0} 2r(2g+r^2 f)dr}dr}}. \tag{4.29}$$

For $r^2 - s^2 > 0$, the characteristic curve through a given point (r, s) crosses the (r_0, s) axis at (r_0, c) with

$$\begin{aligned} c &= \sqrt{r_0^2 + \left[\int_{r_0} 2rfe^{\int_{r_0} 2r(2g+r^2 f)dr}dr - \frac{e^{\int_{r_0} 2r(2g+r^2 f)dr}}{r^2 - s^2}\right]^{-1}} \\ &= \sqrt{r_0^2 + \frac{r^2 - s^2}{(r^2 - s^2)\int_{r_0} 2rfe^{\int_{r_0} 2r(2g+r^2 f)dr}dr - e^{\int 2r(2g+r^2 f)dr}}}. \end{aligned}$$

From equation (4.21), with initial condition $\Psi(r_0) = \psi_{r_0}(c)$, we have $\Psi = \psi_{r_0}(c)$. The solution of the initial value problem (4.24) and (4.25) is therefore

$$\psi(r,s) = \psi_{r_0}\left(\sqrt{r_0^2 + \frac{r^2 - s^2}{(r^2 - s^2)\int_{r_0} 2rfe^{\int_{r_0} 2r(2g+r^2f)dr}dr - e^{\int 2r(2g+r^2f)dr}}}\right). \tag{4.30}$$

Note that, φ given by (4.15) is a solution of equation (4.22), therefore any differentiable real function η of (4.15) is the general solution of equation (4.22). It follows from (4.23) that

$$\phi - s\phi_s = \eta\left(\frac{r^2 - s^2}{(r^2 - s^2)\int 2rfe^{\int 2r(2g+r^2f)dr}dr - e^{\int 2r(2g+r^2f)dr}}\right)/\sqrt{r^2 - s^2}.$$

Therefore (4.14) is the general solution of (4.12). Now the description of the spherically symmetric Douglas metrics follows from Theorem 4.1.

We observe that C.Yu and H. Zhu, [82] gave necessary and sufficient conditions for $F = \alpha\phi(||\beta_x||_\alpha, \frac{\beta}{\alpha})$ to be a Finsler metric for any α and β with $||\beta_x||_\alpha < b_0$. In particular, considering $F(x,y) = |y|\phi(|x|, \frac{\langle x,y\rangle}{|y|})$, then F is a Finsler metric if, and only if, the positive function ϕ satisfies

$$\phi(s) - s\phi_s(s) + (r^2 - s^2)\phi_{ss}(s) > 0, \qquad \text{when } n \geq 2,$$

with the additional inequality

$$\phi(s) - s\phi_s(s) > 0, \qquad \text{when } n \geq 3.$$

Therefore, when ϕ is given by (4.14), F defines a Finsler metric if, and only if, the inequalities (4.16) and (4.17) hold.

4.4 New Families of Douglas Metrics

In this section, we obtain several new families of spherically symmetric Douglas metrics as corollaries of Theorem 4.2. By considering in Theorem 4.2 $f = 0$ and $\eta(\varphi) = A\sqrt{r^2 - s^2}e^{-\int 2rg(r)dr}$, $A > 0$. we obtain the following

Corollary 4.1 *Let $g(r)$ be a differentiable function such that $\int 2rg(r)dr$ is well defined, and let $\phi(r,s)$ be a positive function given by*

$$\phi(r,s) = sh(r) + Ae^{-\int 2rg(r)dr}$$

where $A > 0$ and $h(r)$ is any differentiable function. Then the Finsler metric

$$F(x,y) = |y|\phi\left(|x|, \frac{\langle x, y\rangle}{|y|}\right)$$

is a spherically symmetric Douglas metric defined on $\mathbb{B}^n(r_\mu)$.

In particular, if we choose $g(r) = \frac{r^2}{2}$, we have that $\phi = sh(r) + Ae^{-r^4/4}$, where $A > 0$. Hence, for any differentiable function h of $|x|$ such that $\phi > 0$, we have the following Douglas metric

$$F(x, y) = \langle x, y\rangle h(|x|) + A|y|e^{-\frac{|x|^4}{4}}.$$

Another family of Douglas metrics is obtained from Theorem 4.2, by considering $f(r) = g(r) = 0$, hence $\varphi = -(r^2 - s^2)$ and

$$\phi(r, s) = sh(r) + \frac{\eta(\varphi(r, s))}{\sqrt{r^2 - s^2}} - s\int \frac{1}{s}\frac{\partial}{\partial s}\left(\frac{\eta(\varphi(r, s))}{\sqrt{r^2 - s^2}}\right) ds.$$

By choosing $\eta(\varphi) = \sqrt{-\varphi}\left(\frac{-(\mu+1)\varphi+1}{(-\mu\varphi+1)^{3/2}}\right)$, we have:

Corollary 4.2 *Let* $\phi(r, s)$ *be a function defined by*

$$\phi(r, s) = sh(r) + \frac{[1 + (1 + \mu)r^2][1 + \mu(r^2 - s^2)] + s^2}{\sqrt{1 + \mu(r^2 - s^2)}(1 + \mu r^2)^2}$$

where $\mu \in \mathbb{R}$, *and* $h(r)$ *is any function such that* $\phi(r, s)$ *is positive. Then the following Finsler metric on* $\mathbb{B}^n(r_\mu) \subset \mathbb{R}^n$, *where* $r_\mu = 1/\sqrt{-\mu}$ *if* $\mu < 0$

$$F(x, y) := |y|\phi\left(|x|, \frac{\langle x, y\rangle}{|y|}\right)$$

is a spherically symmetric Douglas metric.

In particular, by choosing in Corollary 4.2

$$h(r) = \frac{2\sqrt{1 + (\mu + 1)r^2}}{(1 + \mu r^2)^2}$$

we obtain the Example given in [82], namely

$$F(x, y) = \frac{(\sqrt{1 + (1 + \mu)|x|^2}\sqrt{(1 + \mu|x|^2)|y|^2 - \mu\langle x, y\rangle^2} + \langle x, y\rangle)^2}{(1 + \mu|x|^2)^2\sqrt{(1 + \mu|x|^2)|y|^2 - \mu\langle x, y\rangle^2}}.$$

As a consequence of Theorem 4.2, for $f(r) = 0$ and $2g(r) = \frac{\zeta\varepsilon+\kappa^2}{(\zeta\varepsilon+\kappa^2)r^2+\varepsilon}$, by choosing $\eta(\varphi) = \varepsilon\sqrt{-(\frac{1}{\varphi} + \kappa^2)^{-1}}$, we get the following result

Corollary 4.3 *Let $\phi(r, s)$ be a function defined by*

$$\phi(r, s) = sh(r) + \frac{\sqrt{\zeta\varepsilon r^2 + \kappa^2 s^2 + \varepsilon}}{\zeta r^2 + 1}$$

where $\zeta, \varepsilon, \kappa$ are any constant real values such that $(\zeta\varepsilon + \kappa^2)r^2 + \varepsilon > 0$, and $h(r)$ is any function such that $\phi(r, s)$ is positive. Then the following spherically symmetric Finsler metric on $\mathbb{B}^n(r_\zeta) \subset \mathbb{R}^n$, where $r_\zeta = 1/\sqrt{-\zeta}$ if $\zeta < 0$,

$$F(x, y) := |y|\phi\left(|x|, \frac{\langle x, y\rangle}{|y|}\right)$$

is of Douglas type.

In particular, when $h(r) = \frac{\kappa}{1+\zeta r^2}$ we have

$$F(x, y) = \frac{\sqrt{\kappa^2\langle x, y\rangle^2 + \varepsilon|y|^2(1 + \zeta|x|^2)}}{1 + \zeta|x|^2} + \frac{\kappa\langle x, y\rangle}{1 + \zeta|x|^2}.$$

When $\kappa = \pm 1$, $\zeta = -1$ and $\varepsilon = 1$, $F(x, y)$ is the Funk metric [22].

In our next application of Theorem 4.2, we do not require f to be zero, in contrast to the previous results. We introduce the following notation

$$I(r) = e^{\int 2r(2g(r)+r^2 f(r))dr}, \tag{4.31}$$

$$L(r, s) = I(r) - (r^2 - s^2)\int 2rf(r)I(r)dr, \tag{4.32}$$

$$T(r) = I(r) - r^2\int 2rf(r)I(r)dr. \tag{4.33}$$

Corollary 4.4 *Let $\phi(r, s)$ be a function defined by*

$$\phi(r, s) = sh(r) + \kappa\frac{\sqrt{L(r, s)}}{T(r)}$$

where $L(r, s)$ and $T(r)$ are defined in (4.32) and (4.33), $\kappa \in \mathbb{R}^+$ and $h(r)$ is any function such that $\phi(r, s)$ is positive. Then, for r and s given by (3.2), the following spherically symmetric Finsler metric on $\mathbb{B}^n(r_\mu) \subset \mathbb{R}^n$,

$$F(x, y) := |y|\phi\left(|x|, \frac{\langle x, y\rangle}{|y|}\right)$$

is a Douglas metric.

Proof It follows from (4.32) that (4.15) can be written as $\varphi(r,s) = \frac{r^2-s^2}{-L(r,s)}$. Moreover, a straightforward computation shows that

$$\frac{\partial}{\partial s}\left(\frac{\sqrt{L(r,s)}}{sT(r)}\right) = -\frac{1}{s^2\sqrt{L(r,s)}}.$$

By choosing $\eta(\varphi) = K\sqrt{-\varphi}$, where $K > 0$ is a constant, it follows from (4.14) that

$$\phi(r,s) = sh(r) + K\frac{\sqrt{L(r,s)}}{T(r)}.$$

Since

$$\frac{-\sqrt{r^2-s^2}}{s}\frac{\partial\eta}{\partial s} = \frac{KI(r)}{(L(r,s))^{\frac{3}{2}}} > 0 \quad \text{and} \quad \frac{\eta}{\sqrt{r^2-s^2}} = \frac{K}{(L(r,s))^{\frac{1}{2}}} > 0,$$

we conclude the proof as a consequence of Theorem 4.2. □

In particular, when we choose $f = c$ a nonzero constant and $2g(r) = (\nu^2 - r^2)c$, when ν is a positive constant, we have

$$I(r) = e^{\nu^2cr^2}, \quad L(r,s) = I(r)(1 - \frac{r^2-s^2}{\nu^2}), \quad T(r) = I(r)(1 - \frac{r^2}{\nu^2})$$

and hence

$$\phi(r,s) = sh(r) + \frac{\kappa\nu\sqrt{\nu^2-(r^2-s^2)}}{e^{\frac{\nu^2cr^2}{2}}(\nu^2-r^2)}$$

where $h(r)$ is any function such that ϕ is positive. It follows that

$$F(x,y) = \langle x, y\rangle h(|x|) + \frac{\kappa\nu\sqrt{(\nu^2-|x|^2)|y|^2+\langle x, y\rangle^2}}{e^{\frac{\nu^2-c|x|^2}{2}}(\nu^2-|x|^2)}$$

is a Douglas metric.

Chapter 5
Projectively Flat Spherically Symmetric Metrics

It has been slightly more than one hundred years since David Hilbert presented a list of 23 outstanding and important problems to the second International Congress of Mathematician in Paris in 1900.

The Hilbert's Fourth Problem is to characterize (not necessary reversible) distance functions on an open subset in $\mathbb{R}^n$ such that straight lines are shortest paths.

Recall that Finsler metric $F = F(x, y)$ on an open subset $\mathcal{U} \subset \mathbb{R}^n$ is said to be *projectively flat* if all geodesics are straight in $\mathcal{U}$.

Distance functions induced by Finsler metrics are regarded as smooth ones. Thus Hilbert's Fourth Problem in the smooth case is to characterize and study projectively flat Finsler metrics on an open subset in $\mathbb{R}^n$. In this chapter, we are going to study and characterize (locally) projectively flat spherically symmetric Finsler metrics.

5.1 Reducible Differential Equation

A Finsler metric $F = F(x, y)$ on an open subset $\mathcal{U} \subset \mathbb{R}^n$ is projectively flat if and only if it satisfies the following system of equation (see Sect. 2.2.1)

$$F_{x^j y^i} y^j = F_{x^i}. \tag{5.1}$$

A function ξ defined on $T\mathcal{U}$ can be expressed as $\xi(x^1, \cdots, x^n, y^1, \cdots, y^n)$. We use the following notation (see proof of Lemma 2.2.3)

$$\xi_0 = \frac{\partial \xi}{\partial x^i} y^i.$$

E. Guo, X. Mo, *The Geometry of Spherically Symmetric Finsler Manifolds*, SpringerBriefs in Mathematics, https://doi.org/10.1007/978-981-13-1598-5_5

By (5.1), we obtain the following

Lemma 5.1 *A Finsler metric* $F = F(x, y)$ *is projectively flat if and only if it satisfies the following system of equations*

$$(F_0)_{y^i} = 2F_{x^i}. \tag{5.2}$$

By using this modified Hamel equation, we have the following (see Corollary 5.2)

Theorem 5.1 *Let* $F = |y|\phi(|x|, \frac{\langle x,y\rangle}{|y|})$ *be a spherically symmetric Finsler metric on* $\mathbb{B}^n(r_\mu)$. *Then* $F = F(x, y)$ *is projectively flat if and only if* $\phi = \phi(r, s)$ *satisfies*

$$s\phi_{rs} + \phi_{ss} - \phi_r = 0. \tag{5.3}$$

Taking $\phi(r, s) = \epsilon + r^\nu f(\frac{s}{r})$ in Theorem 5.1, we have the following

Corollary 5.1 *Let* $F(x, y) := |y|[\epsilon + |x|^\mu f(\frac{\langle x,y\rangle}{|x||y|})]$ *be a spherically symmetric Finsler metric on an open subset* $\mathcal{U} \subset \mathbb{R}^n$. *Then* $F = F(x, y)$ *is projectively flat if and only if*

$$(\lambda^2 - 1)f'' - \mu\lambda f' + \nu f = 0, \tag{5.4}$$

where $\lambda := \frac{\langle x,y\rangle}{|x||y|}$.

Note that (5.4) is the Gegenbauer-type ordinary differential equation [31].

5.2 Solutions of Gegenbauer-Type ODE

In order to find projectively flat spherically symmetric metrics we consider the following ordinary differential equation:

$$\begin{cases} (1-\lambda^2)f''(\lambda) = \mu(f - \lambda f') \\ f(0) = 1, \quad f'(0) = \delta. \end{cases} \tag{5.5}$$

Lemma 5.2 *The solution of (5.5) is*

$$f_\mu(\lambda) = 1 + \delta\lambda + \mu \int_0^\lambda \int_0^\tau (1-\sigma^2)^{\frac{\mu}{2}-1} d\sigma d\tau. \tag{5.6}$$

Furthermore, if $g = g(\lambda)$ *satisfies*

$$\begin{cases} g''(\lambda) = \mu(1-\lambda^2)^{\frac{\mu}{2}-1} \\ g(0) = 1, \quad g'(0) = \delta. \end{cases}$$

Then

$$g = 1 + \delta\lambda + \int_0^\lambda \int_0^\tau (1-\sigma^2)^{\frac{\mu}{2}-1} d\sigma d\tau.$$

Proof When $\mu = 0$, our conclusion is obvious. We assume that $\mu \neq 0$. We have the expansion

$$(1+x)^\xi = \sum_{k=0}^\infty C_\xi^k x^k, \quad x \in (-1,\, 1)$$

where

$$C_\xi^k := \frac{\xi(\xi-1)\cdots(\xi-k+1)}{k!}.$$

By simple calculations, we have

$$\frac{\xi}{k+1} C_{\xi-1}^k = C_\xi^{k+1}, \quad f_\mu(\lambda) = 1 + \delta\lambda + \sum_{k=0}^\infty \frac{\mu C_{\xi-1}^k t^k \lambda^{2k+2}}{(2k+1)(2k+2)},$$

$$f'_\mu(\lambda) = \delta + \sum_{k=0}^\infty \frac{\mu C_{\xi-1}^k t^k \lambda^{2k+1}}{2k+1}, \quad f''_\mu(\lambda) = \mu(1-\lambda^2)^{\frac{\mu}{2}-1}$$

where $\xi = \frac{\mu}{2},\ t = -1$. Thus we have

$$f_\mu(0) = 1, \qquad f'_\mu(0) = \delta$$

and

$$\begin{aligned} f_\mu(\lambda) - \lambda f'_\mu(\lambda) &= 1 + \sum_{k=0}^\infty \mu \left[\frac{C_{\xi-1}^k t^k}{(2k+1)(2k+2)} - \frac{C_{\xi-1}^k t^k}{2k+1} \right] \lambda^{2k+2} \\ &= 1 + \sum_{k=0}^\infty \frac{\xi C_{\xi-1}^k t^{k+1} \lambda^{2k+2}}{k+1} = 1 + \sum_{j=1}^\infty C_\xi^j (t\lambda^2)^j \\ &= (1+t\lambda^2)^\xi = \left(1-\lambda^2\right)^{\frac{\mu}{2}} \\ &= (\tfrac{1}{\mu} - \tfrac{1}{\mu}\lambda^2)\mu(1-\lambda^2)^{\frac{\mu}{2}-1} = (\tfrac{1}{\mu} - \tfrac{1}{\mu}\lambda^2) f''_\mu(\lambda). \end{aligned}$$

It follows that f_μ satisfies (5.5).

Lemma 5.3 *Suppose that f is given in Lemma 5.2. Then*

$$\phi(\rho,\, s) = \epsilon + \rho^\mu f\left(\frac{s}{\rho}\right), \quad \mu \geq 0$$

satisfies

$$\phi(s) - s\phi_s(s) > 0, \quad \phi(s) - s\phi_s(s) + (\rho^2 - s^2)\phi_{ss}(s) > 0, \quad |s| \leq \rho.$$

Proof Direct computations yield

$$\phi_s = \rho^{\mu-1} f'\left(\frac{s}{\rho}\right), \quad \phi_{ss} = \rho^{\mu-2} f''\left(\frac{s}{\rho}\right).$$

It follows that

$$\phi(s) - s\phi_s(s) = \epsilon + \rho^{\mu-1}(\rho f - s f') = \epsilon + \rho^{\mu}(f - \lambda f') \tag{5.7}$$

where we used $\lambda = \frac{s}{\rho}$. Similarly, we get

$$\phi(s) - s\phi_s(s) + (\rho^2 - s^2)\phi_{ss}(s) = \epsilon + \rho^{\mu}(f - \lambda f') + \rho^{\mu}(1 - \lambda^2) f''. \tag{5.8}$$

Assume that $\mu = 0$. In this case

$$f(\lambda) = 1 + \delta\lambda.$$

Then

$$\phi(s) - s\phi_s(s) = \phi(s) - s\phi_s(s) + (\rho^2 - s^2)\phi_{ss}(s) = 1 + \epsilon > 0$$

from (5.7) and (5.8).

Assume that $\mu > 0$. Direct computations yield (cf. proof of Lemma 2.2 in [55])

$$f(\lambda) - \lambda f'(\lambda) = \frac{1}{\mu}(1 - \lambda^2)^{\frac{\mu}{2}}, \quad f''(\lambda) = \mu(1 - \lambda^2)^{\frac{\mu}{2}-1}.$$

It follows that

$$\phi(s) - s\phi_s(s) = \epsilon + \rho^{\mu}(1 - \lambda^2)^{\frac{\mu}{2}} = \epsilon + (\rho^2 - s^2)^{\frac{\mu}{2}} \geq \epsilon > 0, \quad |s| \leq \rho$$

and

$$\begin{aligned}\phi(s) - s\phi_s(s) + (\rho^2 - s^2)\phi_{ss}(s) &= \epsilon + (1 + \mu)\rho^{\mu}(1 - \lambda^2)^{\frac{\mu}{2}} \\ &= \epsilon + (1 + \mu)(\rho^2 - s^2)^{\frac{\mu}{2}} \geq \epsilon > 0, \quad |s| \leq \rho.\end{aligned}$$

Lemma 5.4 *If* $\mu = 2n$, $n \in \mathbb{N}$, *then the solution of* (5.5) *is*

$$f(\lambda) = 1 + \delta\lambda + 2n \sum_{k=0}^{n-1} \frac{(-1)^k C_{n-1}^k \lambda^{2k+2}}{(2k+1)(2k+2)}.$$

Proof Similar to the proof of Lemma 4.1 in [55].

Proposition 5.1 *Let $f(\lambda)$ be a polynomial function defined by*

$$f(\lambda) = 1 + \delta\lambda + 2n \sum_{k=0}^{n-1} \frac{(-1)^k C_{n-1}^k \lambda^{2k+2}}{(2k+1)(2k+2)}.$$

Then the following spherically symmetric metric on an the open subset at origin in $\mathbb{R}^n \setminus \{0\}$

$$F = |y| \left\{ \epsilon + |x|^{2n} f\left(\frac{\langle x, y \rangle}{|x||y|} \right) \right\}$$

is projectively flat.

Proof Combine Corollary 5.1, Lemmas 5.3 and 5.4.

Remark When $\delta = 0$, then

$$f(\lambda) = (2n+1) H_n(-\lambda^2)$$

where

$$H_n(z) := \sum_{i=0}^{k} \left(\frac{1}{2k+1} C_{k+1}^i - \frac{1}{2i-1} C_k^{i-1} \right) z^i.$$

These spherically symmetric metrics, up to a scaling, were constructed in [24, Page 70, Example 4.48].

Proposition 5.2 *Let $f(\lambda)$ be a function defined by*

$$f(\lambda) = \delta\lambda + \frac{(2n-1)!!}{(2n-2)!!} \left[\sqrt{1-\lambda^2} + \lambda \arcsin \lambda - \sum_{k=1}^{n-1} \frac{(2k-2)!!}{(2k+1)!!} (1-\lambda^2)^{\frac{2k+1}{2}} \right].$$

Then the following spherically symmetric metric on an open subset in $\mathbb{R}^n \setminus \{0\}$

$$F = |y| \left\{ \epsilon + |x|^{2n-1} f\left(\frac{\langle x, y \rangle}{|x||y|} \right) \right\}$$

is projectively flat.

Proof Lemma 5.3 tells us F is a spherically symmetric metric. Similar to proofs of Lemma 5.5 and Theorem 5.4 in [55] where we take $\mu = 2n - 1$ in Lemma 5.2.

5.3 Projectively Flat Finsler Metrics in Terms of Hypergeometric Functions

In this section, we are going to give the solutions of (5.5) in terms of hypergeometric functions and manufacture new projectively flat Finsler metrics.

Lemma 5.5 *Let μ be a real number. Then*

$$\int_0^\tau (1-x^2)^{\frac{\mu}{2}-1}dx = \tau \text{hypergeom}\left(\left[\frac{1}{2}, 1-\frac{\mu}{2}\right], \left[\frac{3}{2}\right], \tau^2\right).$$

Proof Let

$$(\lambda)_n = \begin{cases} 1 & \text{if } n=0 \\ \lambda(\lambda+1)\cdots(\lambda+n-1) & \text{if } n \geq 1 \end{cases} \tag{5.9}$$

for $\lambda \in \mathbb{R}$. The *hypergeometric function* is defined by

$$\text{hypergeom}([a, b], c, t) := \sum_{n=0}^{\infty} \frac{(a)_n (b)_n}{n!(c)_n} t^n. \tag{5.10}$$

In particular,

$$\text{hypergeom}([-a, b], b, -t) = (1+t)^a. \tag{5.11}$$

where $a, b \in \mathbb{R}$. From (5.11) and (5.10) we have

$$\begin{aligned} \int_0^\tau (1-x^2)^{\frac{\mu}{2}-1}dx &= \int_0^\tau \sum_{n=0}^{\infty} \frac{(1-\frac{\mu}{2})_n}{n!} x^{2n} dx = \sum_{n=0}^{\infty} \frac{(1-\frac{\mu}{2})_n}{n!} \int_0^\tau x^{2n} dx \\ &= \sum_{n=0}^{\infty} \frac{(1-\frac{\mu}{2})_n}{n!} \frac{\tau^{2n+1}}{2n+1} = \tau \sum_{n=0}^{\infty} \frac{(1-\frac{\mu}{2})_n}{n!} \frac{\tau^{2n}}{2n+1}. \end{aligned} \tag{5.12}$$

By using (5.9), we have

$$\frac{\left(\frac{1}{2}\right)_n}{\left(\frac{3}{2}\right)_n} = \frac{\frac{1}{2}(\frac{1}{2}+1)\cdots(\frac{1}{2}+n-1)}{\frac{3}{2}(\frac{3}{2}+1)\cdots(\frac{3}{2}+n-1)} = \frac{\frac{1}{2}}{\frac{3}{2}+n-1} = \frac{1}{2n+1}.$$

Plugging this into (5.12) yields

$$\begin{aligned}
\int_0^\tau (1-x^2)^{\frac{\mu}{2}-1}dx &= \tau \sum_{n=0}^{\infty} \frac{(1-\frac{\mu}{2})_n}{n!} \frac{\left(\frac{1}{2}\right)_n}{\left(\frac{3}{2}\right)_n} (\tau^2)^n \\
&= \tau \sum_{n=0}^{\infty} \frac{\left(\frac{1}{2}\right)_n (1-\frac{\mu}{2})_n}{n!\left(\frac{3}{2}\right)_n} (\tau^2)^n \\
&= \tau \, \mathrm{hypergeom}\left(\left[\frac{1}{2}, 1-\frac{\mu}{2}\right], \left[\frac{3}{2}\right], \tau^2\right).
\end{aligned} \tag{5.13}$$

Lemma 5.6 *Let μ be a non-zero constant. Then*

$$\begin{aligned}
&\int_0^\lambda \tau \, \mathrm{hypergeom}\left(\left[\frac{1}{2}, 1-\frac{\mu}{2}\right], \left[\frac{3}{2}\right], \tau^2\right) d\tau \\
&\qquad = -\frac{1}{\mu} + \frac{1}{\mu} \mathrm{hypergeom}\left(\left[-\frac{1}{2}, -\frac{\mu}{2}\right], \left[\frac{1}{2}\right], \lambda^2\right).
\end{aligned}$$

Proof By (5.13) we obtain

$$\begin{aligned}
\int_0^\lambda \tau \, \mathrm{hypergeom}\left(\left[\frac{1}{2}, 1-\frac{\mu}{2}\right], \left[\frac{3}{2}\right], \tau^2\right) d\tau &= \int_0^\lambda \sum_{n=0}^{\infty} \frac{\left(\frac{1}{2}\right)_n (1-\frac{\mu}{2})_n}{n!\left(\frac{3}{2}\right)_n} \tau^{2n+1} d\tau \\
&= \sum_{n=0}^{\infty} \frac{\left(\frac{1}{2}\right)_n (1-\frac{\mu}{2})_n}{n!\left(\frac{3}{2}\right)_n} \int_0^\lambda \tau^{2n+1} d\tau \\
&= \sum_{n=0}^{\infty} \frac{\left(\frac{1}{2}\right)_n (1-\frac{\mu}{2})_n}{n!\left(\frac{3}{2}\right)_n} \frac{\lambda^{2n+2}}{2n+2}.
\end{aligned}$$

Taking $m = n + 1$ we obtain

$$\begin{aligned}
&\int_0^\lambda \tau \, \mathrm{hypergeom}\left(\left[\frac{1}{2}, 1-\frac{\mu}{2}\right], \left[\frac{3}{2}\right], \tau^2\right) d\tau \\
&\qquad = \frac{1}{2} \sum_{m=1}^{\infty} \frac{\left(\frac{1}{2}\right)_{m-1} (1-\frac{\mu}{2})_{m-1}}{m!\left(\frac{3}{2}\right)_{m-1}} \left(\lambda^2\right)^m.
\end{aligned} \tag{5.14}$$

By straightforward computations one obtains

$$\left(\frac{1}{2}\right)_{m-1}=-2\times\left(-\frac{1}{2}\right)_m,\quad \left(1-\frac{\mu}{2}\right)_{m-1}=-\frac{2}{\mu}\left(-\frac{\mu}{2}\right)_m,$$
$$\left(\frac{3}{2}\right)_{m-1}=2\times\left(\frac{1}{2}\right)_m. \tag{5.15}$$

Substituting (5.15) into (5.14) yields

$$\begin{aligned}
&\int_0^\lambda \tau\,\text{hypergeom}\left(\left[\frac{1}{2},1-\frac{\mu}{2}\right],\left[\frac{3}{2}\right],\tau^2\right)d\tau\\
&=\frac{1}{2}\sum_{m=1}^{\infty}\frac{-2\left(-\frac{1}{2}\right)_m\left(-\frac{2}{\mu}\right)(-\frac{\mu}{2})_m}{m!2\left(\frac{1}{2}\right)_m}\left(\lambda^2\right)^m\\
&=\frac{1}{\mu}\sum_{m=1}^{\infty}\frac{\left(-\frac{1}{2}\right)_m(-\frac{\mu}{2})_m}{m!\left(\frac{1}{2}\right)_m}\left(\lambda^2\right)^m\\
&=-\frac{1}{\mu}+\frac{1}{\mu}\text{hypergeom}\left(\left[-\frac{1}{2},-\frac{\mu}{2}\right],\left[\frac{1}{2}\right],\lambda^2\right).
\end{aligned}$$

Theorem 5.2 *Let $f(\lambda)$ be a function defined by*

$$f(\lambda)=\delta\lambda+\text{hypergeom}\left(\left[-\frac{1}{2},-\frac{\mu}{2}\right],\left[\frac{1}{2}\right],\lambda^2\right)$$

where δ and μ are constants ($\mu\geq 0$). Then the following spherically symmetric metric on an open subset in $\mathbb{R}^n\backslash\{0\}$

$$F=|y|\left\{\epsilon+|x|^\mu f\left(\frac{\langle x,y\rangle}{|x||y|}\right)\right\}$$

is projectively flat.

Proof From Lemma 5.2, the solution of (5.5) is

$$f(\lambda)=1+\delta\lambda+\mu\int_0^\lambda\int_0^\tau(1-\sigma^2)^{\frac{\mu}{2}-1}d\sigma d\tau.$$

Combining this with Lemmas 5.5 and 5.6 we obtain

$$
\begin{aligned}
f(\lambda) &= 1 + \delta\lambda + \mu \int_0^{\lambda} \tau \,\text{hypergeom}\left(\left[\frac{1}{2}, 1 - \frac{\mu}{2}\right], \left[\frac{3}{2}\right], \tau^2\right) d\tau \\
&= 1 + \delta\lambda + \mu\left[-\frac{1}{\mu} + \frac{1}{\mu}\text{hypergeom}\left(\left[-\frac{1}{2}, -\frac{\mu}{2}\right], \left[\frac{1}{2}\right], \lambda^2\right)\right] \\
&= \delta\lambda + \text{hypergeom}\left(\left[-\frac{1}{2}, -\frac{\mu}{2}\right], \left[\frac{1}{2}\right], \lambda^2\right)
\end{aligned}
\tag{5.16}
$$

for arbitrary nonzero μ. When $\mu = 0$, (5.16) is automatically true. Lemma 5.3 tells us F is that a spherically symmetric metric when $\mu \geq 0$. By Corollary 5.1, the spherically symmetric metric

$$
F = |y| \left\{ \epsilon + |x|^{\mu} f\left(\frac{\langle x, y \rangle}{|x||y|} \right) \right\}
$$

is projectively flat.

5.4 Projectively Flat Finsler Metrics in Terms of Error Functions

In this section we are going to find the general solution ϕ of (5.3). Then we give a lot of new projectively flat spherically symmetric metrics of in terms of error functions.

Proposition 5.3 *For $s > 0$, the general solution ϕ of* (5.3) *is given by*

$$
\phi(r, s) = sg(r) - s \int_{s_0}^{s} t^{-2} f(r^2 - t^2) dt \tag{5.17}
$$

where $s_0 \in (0, s]$.

Proof Note that $s > 0$. We obtain that (5.3) is equivalent to the following

$$
sz_r + rz_s = 0 \tag{5.18}
$$

where

$$
z := \phi - s\phi_s. \tag{5.19}
$$

The characteristic equation of quasi-linear PDE (5.18) is

$$\frac{dr}{s} = \frac{ds}{r} = \frac{dz}{0}. \tag{5.20}$$

It follows that

$$r^2 - s^2 = c_1, \quad z = c_2$$

are independent integrals of (5.20). Hence the solution of (5.18) is

$$z = f(r^2 - s^2) \tag{5.21}$$

where f is a continuously differentiable function. Hence we have

$$\phi - s\phi_s = f(r^2 - s^2) \tag{5.22}$$

It follows that the solution of (5.3) satisfies (5.22). Conversely, suppose that (5.22) holds. Then we obtain (5.18) and (5.19). Thus ϕ satisfies (5.3). We conclude that (5.22) and (5.3) are equivalent.

Now we consider $s \in [s_0, +\infty)$ where $s_0 > 0$. Put

$$\phi = s\psi. \tag{5.23}$$

It follows that $\phi_s = \psi + s\psi_s$. Together with (5.22) yields

$$f(r^2 - s^2) = s\psi - s(\psi + s\psi_s) = -s^2\psi_s.$$

Thus

$$\psi = g(r) - \int_{s_0}^{s} t^{-2} f(r^2 - t^2)dt.$$

Plugging this into (5.23) yields (5.17).

Remark Similarly, we have the general solution of (5.3) for $s < 0$.

In mathematics, the *error function* is a special function (non-elementary) of sigmoid shape which occurs in probability, statistics and partial differential equations[2, 40]. It is defined by $\mathrm{erf}(x) := \frac{2}{\sqrt{\pi}} \int_0^x e^{-t^2} dt$.

Now we manufacture projectively flat spherically symmetric metrics in terms of error functions.

Taking $f(u) = e^{\lambda u}$ in (5.17) where $\lambda \in \mathbb{R}^+$ we have

$$\begin{aligned}
\int t^{-2} f(r^2 - t^2) dt &= \int t^{-2} e^{\lambda(r^2 - t^2)} dt \\
&= e^{\lambda r^2} \int t^{-2} e^{-\lambda t^2} dt \\
&= e^{\lambda r^2} \left(- \int e^{-\lambda t^2} dt^{-1} \right) \\
&= e^{\lambda r^2} \left(\int t^{-1} de^{-\lambda t^2} - t^{-1} e^{-\lambda t^2} \right) \\
&= -e^{\lambda r^2} \left(t^{-1} e^{-\lambda t^2} + 2\lambda \int e^{-\lambda t^2} dt \right).
\end{aligned}$$

Combining with (5.17) we have

$$\begin{aligned}
\phi(r, s) &= sg(r) + se^{\lambda r^2} \left(t^{-1} e^{-\lambda t^2} |_{s_0}^{s} + 2\lambda \int_{s_0}^{s} e^{-\lambda t^2} dt \right) \\
&= sg_1(r) + e^{\lambda r^2} \left(e^{-\lambda s^2} + 2\lambda s \int_{s_0}^{s} e^{-\lambda t^2} dt \right).
\end{aligned} \tag{5.24}$$

On the other hand,

$$\int_0^r e^{-\lambda t^2} dt = \frac{1}{\sqrt{\lambda}} \int_0^{\sqrt{\lambda} r} e^{-\lambda x^2} dx = \frac{\sqrt{\pi}}{2\sqrt{\lambda}} \mathrm{erf}(\sqrt{\lambda}\, r).$$

Substituting this into (5.24) yields

$$\begin{aligned}
\phi(r, s) &= sg_1(r) + e^{\lambda r^2} \left\{ e^{-\lambda s^2} + \sqrt{\lambda \pi} \left[\mathrm{erf}(\sqrt{\lambda}\, s) - \mathrm{erf}(\sqrt{\lambda}\, s_0) \right] s \right\} \\
&= sg_2(r) + e^{\lambda r^2} \left[e^{-\lambda s^2} + \sqrt{\lambda \pi} s \ \mathrm{erf}(\sqrt{\lambda}\, s) \right].
\end{aligned} \tag{5.25}$$

Together with Theorem 5.1 and Proposition 5.3 we obtain (see [45, 74])

Theorem 5.3 *Let $\phi(r, s)$ be a function define by*

$$\phi(r, s) = sg(r) + e^{\lambda r^2} \left[e^{-\lambda s^2} + \sqrt{\lambda \pi} s \ \mathrm{erf}(\sqrt{\lambda}\, s) \right]$$

where $\lambda > 0$, erf(,) denote the error function and g is any function. Then the following spherically symmetric Finsler metric of on an open subset in $\mathbb{R}^n$

$$F = |y| \phi \left(|x|, \frac{\langle x, y \rangle}{|y|} \right)$$

is projectively flat.

5.5 Locally Projectively Flat Finsler Metrics

A Finsler metric F on a manifold M is said to be *locally projectively flat* if at any point, there is a local coordinate system (x^i) in which the geodesics are straight lines as point sets, equivalently, F is pointwise projectively related to a locally Minkowskian metric. For definition of pointwise projectively related see Chap. 8 below. Riemannian metrics of constant (sectional) curvature are locally projectively flat. The converse is also true according to Beltrami's theorem.

In this section we are going to characterize locally projectively flat spherically symmetric metrics.

Theorem 5.4 *Let $F(x, y) = |y|\phi(r, s)$ be a spherically symmetric Finsler metric on $\mathbb{B}^n(r_\mu)$ $(n > 2)$, where $r := |x|$, $s := \frac{\langle x, y\rangle}{|y|}$. Then, the following assertions are equivalent:*

(i) *F is locally projectively flat;*

(ii) *$\phi = \phi(r, s)$ satisfies*

$$\left[(r^2 - s^2)Q - 1\right] r\phi_{ss} - s\phi_{rs} + \phi_r + rQ(\phi - s\phi_s) = 0 \tag{5.26}$$

where $Q = Q(r, s)$ is given by

$$Q(r, s) = f(r) + \frac{2rf(r)^2 + f'(r)}{r + 2f(r)r^3} s^2 \tag{5.27}$$

where $f = f(r)$ is a differential function.

(iii) *$\phi = \phi(r, s)$ satisfies* (5.26) *where $Q = Q(r, s)$ is a polynomial in s and F is of scalar curvature.*

Proof According to Douglas' result, Finsler metric $F(x, y)$ on $\mathbb{B}^n(r_\mu)$ $(n > 2)$ is locally projectively flat if and only if F has vanishing Weyl curvature and Douglas curvature [20]. On the other hand, note that $F = |y|\phi\left(|x|, \frac{\langle x, y\rangle}{|y|}\right)$ is spherically symmetric, we have the following:

(i) F has vanishing Douglas curvature if and only if $\phi = \phi(r, s)$ satisfies (5.26) where (see (4.11))

$$Q(r, s) = f(r) + g(r)s^2. \tag{5.28}$$

(ii) If $\phi = \phi(r, s)$ satisfies (5.26) where $Q = Q(r, s)$ is a polynomial in s. Then F has vanishing Weyl curvature if and only if (5.28) holds and (cf. (8.14)[38], Proposition 3.2 and 4.1)

$$2f^2 + \frac{1}{r}f' - g + 2r^2 fg = 0.$$

Now our theorem is an immediate conclusion of (i), (ii) and Douglas' result. Recall that a Finsler metric has vanishing Weyl curvature if and only if it is of scalar curvature [42].

As a consequence of Theorem 5.4, by taking $f = 0$ in (5.27), we obtain the following result obtained by Huang and the second author (see [25], Theorem 1.1).

Corollary 5.2 *Let* $F = |y|\phi\left(|x|, \frac{\langle x, y\rangle}{|y|}\right)$ *be a spherically symmetric Finsler metric on* $\mathbb{B}^n(\nu)$. *Then* $F = F(x, y)$ *is projectively flat if and only if* $\phi = \phi(r, s)$ *satisfies*

$$s\phi_{rs} + r\phi_{ss} - \phi_r = 0. \tag{5.29}$$

Theorem 5.4 also generalizes a result previously only known in the case of Q being a polynomial in s [86].

Chapter 6
Spherically Symmetric Metrics of Scalar Curvature

In Riemannian geometry, one has the concept of sectional curvature. Its analogue in Finsler geometry is called *flag curvature*. It is one of important problems in Finsler geometry to study and characterize Finsler metrics of scalar (flag) curvature because Finsler metrics of scalar curvature and dimension $m \geqslant 3$ are the natural extension of Riemannian metrics of constant sectional curvature. A Finsler metric F is said to be of *scalar (flag) curvature* if the flag curvature κ at a point x is independent of the tangent plane $P \subseteq T_xM$. In general case, the flag curvature $\kappa = \kappa(P, y)$ is a function of tangent planes $P = \text{span}\{y, v\} \subset T_xM$ and direction $y \in P\backslash\{0\}$.

In this chapter, we find equations that characterize spherically symmetric Finsler metrics of scalar curvature (see Theorem 6.1 below). After investigating these partial differential equations we produce infinitely many non-projectively flat spherically symmetric Finsler metrics of scalar curvature in terms of elementary functions with respect to s.

6.1 Some Lemmas

Let $F = |y|\phi\left(\frac{|x|^2}{2}, \frac{\langle x, y\rangle}{|y|}\right)$ be a spherically symmetric Finsler metric on $\mathbb{B}^m(r_\mu)$. By (2.19), (2.21) and (2.25), the geodesic coefficients G^i can be expressed by (see [50], Definition 3.3.8)

$$G^i := g^{ij}G_j = \frac{r^2}{2}(r^i, s^i)\begin{pmatrix} u \\ v \end{pmatrix}, \tag{6.1}$$

E. Guo, X. Mo, *The Geometry of Spherically Symmetric Finsler Manifolds*, SpringerBriefs in Mathematics, https://doi.org/10.1007/978-981-13-1598-5_6

where

$$v := \frac{\Sigma}{\Delta} = \frac{s\phi_{ts} + \phi_{ss} - \phi_t}{\phi - s\phi_s + (2t - s^2)\phi_{ss}}, \quad \Sigma := s\phi_{ts} + \phi_{ss} - \phi_t. \tag{6.2}$$

and

$$u = \frac{1}{\phi}[\phi_s + s\phi_t - (2t - s^2)\phi_s v].$$

Remark 6.1 By (6.2), we see that F is projectively flat if and only if v vanishes (See Remark 2.1.1).

By using (2.6), (2.7), (2.13), (2.15), (6.1) and Lemma 2.1.1 we obtain

$$(G^i)_{x^j} = c_3\delta^i_j + (r^i, s^i)X_3\begin{pmatrix} r_j \\ s_j \end{pmatrix}, \tag{6.3}$$

where

$$c_3 := \frac{r^2}{2}v, \quad X_3 = \frac{r^2}{2}\begin{pmatrix} u_s + su_t - v & u_t \\ v_s + sv_t & v_t \end{pmatrix}. \tag{6.4}$$

It is easy to see $u = u(t, s)$, $v = v(t, s)$. Together with Lemma 2.1.1 we have $u_{y^j} = r_j u_r + s_j \frac{u_s}{r} = s_j \frac{u_s}{r}$. Similarly, we obtain $v_{y^j} = s_j \frac{v_s}{r}$. The connection coefficients N^i_j satisfy (cf. (2.6) and (2.8) in [17]),

$$N^i_j = \frac{\partial G^i}{\partial y^j} = c_4\delta^i_j + (r^i, s^i)X_4(r_j, s_j)^T, \tag{6.5}$$

where

$$c_4 = \frac{r}{2}(u - sv), \quad X_4 = \frac{r}{2}\begin{pmatrix} u + sv & u_s - v \\ 2v & v_s \end{pmatrix} \tag{6.6}$$

and we have used (6.1) and (2.13).

The *Reeb field* (spray in an alternative terminology in (5.8) in [66]) is defined by

$$\xi = y^i\frac{\partial}{\partial x^i} - 2G^i\frac{\partial}{\partial y^i}, \tag{6.7}$$

where G^i are the geodesic coefficients of F. From (2.13), (2.14), (2.15), (2.19), (6.1) and (6.7), we have the following:

Lemma 6.1 *Let $f = f(r, t, s)$ be a function on a domain $\mathscr{U} \subseteq \mathbb{R}^3$. Then*

$$\xi(f) = r\{sf_t + [1 - (2t - s^2)v]f_s - ruf_r\}.$$

Moreover,

$$(\xi(r_i), \xi(s_i)) = (r_i, s_i)T, \tag{6.8}$$

where

$$T = rv\begin{pmatrix} 0 & 2t - s^2 \\ -1 & s \end{pmatrix}.$$

By using (6.5) and (6.8) we get

$$\xi(N^i_j) = c_5\delta^i_j + (r^i, s^i)X_5\begin{pmatrix} r_j \\ s_j \end{pmatrix},$$

where

$$c_5 = \xi(c_4), \quad X_5 = \xi(X_4) + TX_4 + X_4T'. \tag{6.9}$$

From (6.5) and (2.19), we have

$$N^i_k N^k_j = c_6\delta^i_j + (r^i, s^i)X_6\begin{pmatrix} r_j \\ s_j \end{pmatrix}, \tag{6.10}$$

where

$$c_6 = c_4^2, \quad X_6 = 2c_4X_4 + X_4SX_4. \tag{6.11}$$

The *Riemann curvature* of F is a family of endomorphisms $R_y = R^i_j dx^j \otimes \frac{\partial}{\partial x^i}$: $T_xM \to T_xM$, defined by

$$R^i_j = 2\frac{\partial G^i}{\partial x^j} - y^k\frac{\partial^2 G^i}{\partial x^k \partial y^j} + 2G^k\frac{\partial^2 G^i}{\partial y^k \partial y^j} - \frac{\partial G^i}{\partial y^k}\frac{\partial G^k}{\partial y^j} \tag{6.12}$$

where G^i are the geodesic coefficients of F. Together with (6.5), (6.7), (6.3), (6.9) and (6.10), we have

$$\begin{aligned} R^i_j &= 2(G^i)_{x^j} - (y^k\tfrac{\partial}{\partial x^k} - 2G^k\tfrac{\partial}{\partial y^k})(\tfrac{\partial G^i}{\partial y^j}) - \tfrac{\partial G^i}{\partial y^k}\tfrac{\partial G^k}{\partial y^j} \\ &= 2(G^i)_{x^j} - \xi(N^i_j) - N^i_k N^k_j = c_7\delta^i_j + (r^i, s^i)X_7\begin{pmatrix} r_j \\ s_j \end{pmatrix}, \end{aligned} \tag{6.13}$$

where

$$c_7 = 2c_3 - c_5 - c_6, \quad X_7 = 2X_3 - X_5 - X_6. \tag{6.14}$$

Combining the first formula of (6.14) with (6.4), (6.9), (6.11) and Lemma 6.1, we get

$$\begin{aligned} c_7 &= r^2 v - \xi(c_4) - c_4^2 \\ &= r^2 v - r\left\{s\tfrac{\partial}{\partial t} + [1-(2t-s^2)v]\tfrac{\partial}{\partial s} - ru\tfrac{\partial}{\partial r}\right\}\left[\tfrac{r}{2}(u-sv)\right] - \tfrac{r^2}{4}(u-sv)^2. \end{aligned}$$

where v is given in (6.2). It follows that c_7 is of the form $r^2\psi(t,s)$. From (6.4), (6.14), (6.9) and (6.11) we have

$$X_7 = r^2\begin{pmatrix} u_s + su_t - v & u_t \\ v_s + sv_t & v_t \end{pmatrix} - (I) \tag{6.15}$$

where

$$\begin{aligned} (I) &= X_5 + X_6 \\ &= \xi(X_4) + TX_4 + X_4T' + 2c_4X_4 + X_4SX_4 \\ &= \frac{r}{2}\left\{s\frac{\partial}{\partial t} + [1-(2t-s^2)v]\frac{\partial}{\partial s} - ru\frac{\partial}{\partial r}\right\}\left(r\begin{pmatrix} u+sv & u_s - v \\ 2v & v_s \end{pmatrix}\right) \\ &\quad + \frac{r^2 v}{2}\begin{pmatrix} 0 & 2t-s^2 \\ -1 & s \end{pmatrix}\begin{pmatrix} u+sv & u_s - v \\ 2v & v_s \end{pmatrix} + \frac{r^2 v}{2}\begin{pmatrix} u+sv & u_s - v \\ 2v & v_s \end{pmatrix}\begin{pmatrix} 0 & -1 \\ 2t-s^2 & s \end{pmatrix} \\ &\quad + \frac{r^2}{4}\left\{2(u-sv)I_2 + \begin{pmatrix} u+sv & u_s - v \\ 2v & v_s \end{pmatrix}\begin{pmatrix} 1 & 0 \\ 0 & 2t-s^2 \end{pmatrix}\right\}\begin{pmatrix} u+sv & u_s - v \\ 2v & v_s \end{pmatrix}. \end{aligned} \tag{6.16}$$

We denote X_7 by

$$X_7 := \begin{pmatrix} \gamma_1 & \gamma_2 \\ \gamma_3 & \gamma_4 \end{pmatrix}. \tag{6.17}$$

It is easy to see, from (6.15), (6.16), and (6.17), $\gamma_1, \cdots, \gamma_4$ are of the form $r^2\psi(t,s)$. Summarizing up, we have the following:

Remark 6.2 $c_7, \gamma_1, \cdots, \gamma_4$ are of the form $r^2\psi(t,s)$.

It is known that the Riemann curvature satisfies the following equations ([17], page 42 (2.44))

$$R^i_{\ j} y^j = 0, \qquad R^i_{\ j} F_{y^i} = 0. \tag{6.18}$$

Together with (2.7), (6.13), (6.17) and (2.19), we have

$$0 = R^i_j r^j = c_7 r^i + r^i \gamma_1 + s^i \gamma_3 = (c_7 + \gamma_1) r^i + \gamma_3 s^i.$$

It follows that

$$0 = [(c_7 + \gamma_1) r^i + \gamma_3 s^i] r_i = c_7 + \gamma_1 \tag{6.19}$$

and

$$0 = (\gamma_3 s^i) s_i = \gamma_3 (2t - s^2) = \gamma_3 \frac{|x|^2 |y|^2 - \langle x, y\rangle^2}{|y|^2}.$$

Taking x and y with $x \wedge y \neq 0$ we obtain

$$\gamma_3 = 0. \tag{6.20}$$

Plugging (6.13) and (2.11) into the second equation of (6.18) yields

$$0 = \left[c_7 \delta^i_j + (r^i, s^i) \begin{pmatrix} \gamma_1 & \gamma_2 \\ \gamma_3 & \gamma_4 \end{pmatrix} \begin{pmatrix} r_j \\ s_j \end{pmatrix} \right] (\phi r_i + \phi_s s_i) = [c_7 \phi_s + \gamma_2 \phi + \gamma_4 (2t - s^2) \phi_s] s_j.$$

Thus we have

$$c_7 \phi_s + \gamma_2 \phi + \gamma_4 (2t - s^2) \phi_s = 0. \tag{6.21}$$

6.2 Scalar Curvature Equation

In this section, we are going to find equations that characterize spherically symmetric Finsler metrics of scalar flag curvature.

A Finsler metric F on a manifold M is said to be of *scalar curvature* if the flag curvature $\kappa(P, y) = \kappa(x, y)$ is a scalar function on the slit tangent bundle $TM \backslash \{0\}$. The *angular form* $h_y(u, v) := h_{ij}(y) u^i v^j$ on $T_x M$ is given by

$$h_{ij} := F F_{y^i y^j}.$$

$h = \{h_y\}$ is called the *angular metric*.

Lemma 6.2 *For $m \geq 3$, suppose that the quantities $\hat{R}^i_j$ on $\mathbb{B}^m(r_\mu)$ satisfies*

$$\hat{R}^i_j = \hat{c}_0 \delta^i_j + (r^i, s^i) \begin{pmatrix} \hat{c}_1 & \hat{c}_2 \\ \hat{c}_3 & \hat{c}_4 \end{pmatrix} \begin{pmatrix} r_j \\ s_j \end{pmatrix} \tag{6.22}$$

where $\hat{c}_k = \hat{c}_k(r, t, s)$, $k = 0, 1, \cdots, 4$. *Then* $\hat{R}^i_{\ j} = 0$ *if and only if* $\hat{c}_0 = \hat{c}_1 = \cdots = \hat{c}_4 = 0$.

Proof Taking x and y with $x \wedge y \neq 0$, we obtain

$$\Sigma(r^i)^2 \Sigma(s^i)^2 - (\Sigma r^i s^i)^2 = |y|^2|x|^2 - \langle x, y\rangle^2 > 0. \tag{6.23}$$

It follows that $(r^1, \cdots, r^m)$ and $(s^1, \cdots, s^m)$ are not collinear. Note that $m \geq 3$. Hence there exists $(\omega_1, \cdots, \omega_m) \neq 0$, such that

$$\omega_i r^i = \omega_i s^i = 0. \tag{6.24}$$

Sufficiency is an immediate consequence of (6.22). Conversely, suppose that $\hat{R}^i_{\ j} = 0$. Then

$$0 = \hat{R}^i_{\ j}\omega_i = \left[\hat{c}_0\delta^i_j + (r^i, s^i)\begin{pmatrix} \hat{c}_1 & \hat{c}_2 \\ \hat{c}_3 & \hat{c}_4 \end{pmatrix}\begin{pmatrix} r_j \\ s_j \end{pmatrix}\right]\omega_i = \hat{c}_0\omega_j$$

where we have used (6.22) and (6.24). It follows that $\hat{c}_0 = 0$. Together with (6.22) and (2.19) we have

$$0 = \hat{R}^i_{\ j} r^j = (r^i, s^i)\begin{pmatrix} \hat{c}_1 & \hat{c}_2 \\ \hat{c}_3 & \hat{c}_4 \end{pmatrix}\begin{pmatrix} 1 \\ 0 \end{pmatrix} = r^i\hat{c}_1 + s^i\hat{c}_3.$$

Note that (r^i) and (s^i) are not collinear. Hence we get

$$\hat{c}_1 = \hat{c}_3 = 0. \tag{6.25}$$

Further, from (6.22), (6.24), (6.25) and (2.19), we obtain

$$0 = \hat{R}^i_{\ j} s^j = (r^i, s^i)\begin{pmatrix} 0 & \hat{c}_2 \\ 0 & \hat{c}_4 \end{pmatrix}\begin{pmatrix} r_j \\ s_j \end{pmatrix} s^j = (2t - s^2)(r^i\hat{c}_2 + s^i\hat{c}_4).$$

Taking this together with (6.23) and (2.6) we have $\hat{c}_2 = \hat{c}_4 = 0$.

Proposition 6.1 *Let* $m \geq 3$ *and let* $F = |y|\phi(\frac{|x|^2}{2}, \frac{\langle x, y\rangle}{|y|})$ *be a Finsler metric on* $\mathbb{B}^m(r_\mu)$. *Then* F *is of scalar curvature if and only if* ϕ *satisfies* $\gamma_4 = 0$ *where* γ_4 *is defined in* (6.17).

Proof Let $h^i_{\ j} = g^{ik}h_{kj}$. Then

$$F^2 h^i_{\ j} = F^2 g^{ik}h_{kj} = F^2\delta^i_j - g_{jk}y^k y^i = r^2\phi^2\delta^i_j + r^2(r^i, s^i)\begin{pmatrix} -\phi^2 & -\phi\phi_s \\ 0 & 0 \end{pmatrix}\begin{pmatrix} r_j \\ s_j \end{pmatrix}$$

where we have used (2.19) and Lemma 2.1.2. Together with (6.13) and (6.17), we have

$$R^i_j - \kappa F^2 h^i_j = (c_7 - \kappa r^2\phi^2)\delta^i_j + (r^i, s^i)\begin{pmatrix} \gamma_1 + \kappa r^2\phi^2 & \gamma_2 + \kappa r^2\phi\phi_s \\ \gamma_3 & \gamma_4 \end{pmatrix}\begin{pmatrix} r_j \\ s_j \end{pmatrix} \tag{6.26}$$

where $\kappa = \kappa(x, y)$ is a scalar function on $\mathbb{B}^m(r_\mu)$. F has scalar curvature with flag curvature $\kappa(x, y)$ is equivalent to the following equation (see [17], page 110)

$$R^i_j = \kappa F^2 h^i_j. \tag{6.27}$$

By (6.26) and Lemma 6.2, (6.27) holds if and only if

$$c_7 - \kappa r^2\phi^2 = 0,\ \gamma_1 + \kappa r^2\phi^2 = 0,\ \gamma_2 + \kappa r^2\phi\phi_s = 0,\ \gamma_3 = 0,\ \gamma_4 = 0. \tag{6.28}$$

By (6.20) and (6.19), (6.28) holds if and only if

$$c_7 = \kappa r^2\phi^2,\ \gamma_2 = -\kappa r^2\phi\phi_s,\ \gamma_4 = 0. \tag{6.29}$$

Now the necessary condition is obvious. Conversely, we suppose that $\gamma_4 = 0$. Plugging this into (6.21) yields

$$c_7\phi_s + \gamma_2\phi = 0. \tag{6.30}$$

By using Remark 6.2, we express c_7 by

$$c_7 = r^2\lambda(t, s).$$

We put $\kappa(t, s) = \frac{\lambda(t,s)}{[\phi(t,s)]^2}$. Together with (2.6) we have $\kappa = \kappa(x, y)$ is a scalar function on the slit tangent bundle $T\mathbb{B}^m(r_\mu)\backslash\{0\}$. Furthermore, $c_7 = \kappa r^2\phi^2$. Plugging this into (6.30) yields $\kappa r^2\phi^2\phi_s + \gamma_2\phi = 0$. It follows that

$$\gamma_2 = -\kappa r^2\phi\phi_s.$$

Thus we have (6.29), therefore F is of scalar curvature with flag curvature κ.

Proposition 6.2 *For a spherically symmetric Finsler metric* $F = |y|\phi(\frac{|x|^2}{2}, \frac{\langle x,y\rangle}{|y|})$,

$$\gamma_4 = -\frac{r^2}{4}[(2t - s^2)(v_s^2 - 2vv_{ss}) + 2(sv_{ts} + v_{ss}) + 2svv_s - 4(v_t + v^2)]. \tag{6.31}$$

Proof It follows from (6.15), (6.16) and (6.17) that

$$\begin{aligned}\begin{pmatrix}\gamma_1 & \gamma_2\\ \gamma_3 & \gamma_4\end{pmatrix} = & r^2\begin{pmatrix}u_s+su_t-v & u_t\\ v_s+sv_t & v_t\end{pmatrix}\\ & -\frac{r}{2}\left\{s\frac{\partial}{\partial t}+[1-(2t-s^2)v]\frac{\partial}{\partial s}-ru\frac{\partial}{\partial r}\right\}\left(r\begin{pmatrix}u+sv & u_s-v\\ 2v & v_s\end{pmatrix}\right)\\ & -\frac{r^2 v}{2}\begin{pmatrix}0 & 2t-s^2\\ -1 & s\end{pmatrix}\begin{pmatrix}u+sv & u_s-v\\ 2v & v_s\end{pmatrix}\\ & -\frac{r^2 v}{2}\begin{pmatrix}u+sv & u_s-v\\ 2v & v_s\end{pmatrix}\begin{pmatrix}0 & -1\\ 2t-s^2 & s\end{pmatrix}\\ & -\frac{r^2}{4}\times(I)\times\begin{pmatrix}u+sv & u_s-v\\ 2v & v_s\end{pmatrix}\end{aligned} \tag{6.32}$$

where

$$\begin{aligned}(I):&=2(u-sv)I_2+\begin{pmatrix}u+sv & u_s-v\\ 2v & v_s\end{pmatrix}\begin{pmatrix}1 & 0\\ 0 & 2t-s^2\end{pmatrix}\\ &=\begin{pmatrix}3u-sv & (2t-s^2)(u_s-v)\\ 2v & 2(u-sv)+(2t-s^2)v_s\end{pmatrix}.\end{aligned} \tag{6.33}$$

Plugging (6.33) into (6.32) yields

$$\begin{aligned}\gamma_4 = & r^2 v_t - \tfrac{r}{2}\{s\tfrac{\partial}{\partial t}+[1-(2t-s^2)v]\tfrac{\partial}{\partial s}-ru\tfrac{\partial}{\partial r}\}(rv_s)\\ & -\tfrac{r^2 v}{2}[-(u_s-v)+sv_s-2v+sv_s]\\ & -\tfrac{r^2}{4}\{2v(u_s-v)+v_s[2(u-sv)+(2t-s^2)v_s]\}.\end{aligned}$$

Then (6.31) holds.

By combining Proposition 6.1 with Proposition 6.2, we have

Theorem 6.1 *Let $m \geq 3$ and let $F = |y|\phi(\frac{|x|^2}{2}, \frac{\langle x,y\rangle}{|y|})$ be a spherically symmetric Finsler metric on $\mathbb{B}^m(r_\mu)$. Then F is of scalar curvature if and only if ϕ satisfies*

$$(2t-s^2)(v_s^2-2vv_{ss})+2(sv_{ts}+v_{ss})+2svv_s-4(v_t+v^2)=0 \tag{6.34}$$

where v satisfies (6.2) *(also see* (6.38) *below).*

6.3 Non-projectively Flat Spherically Symmetric Metrics of Scalar Curvature

Equation (6.34) has a trivial solution $v = 0$. In this case, F is projectively flat (see Remark 6.1). It follows that the non-trivial solutions of (6.34) produce non-projectively flat Finsler metrics of scalar curvature in terms of Remark 6.1. Let us consider the special solution of (6.34) in the form $v = v(t)$. In this case, $v_s = 0$. Plugging this into (6.34) yields

$$v_t + v^2 = 0. \tag{6.35}$$

The solution of (6.35) for v is given by

$$v = \frac{1}{t+c}, \qquad v = 0 \tag{6.36}$$

where $c = constant$. We rewrite (6.36) as follows

$$v = \frac{\lambda}{1+\lambda t}, \qquad v = \frac{1}{t} \tag{6.37}$$

where $\lambda = constant$. By the first equation of (6.2), we have

$$s\phi_{ts} + \phi_{ss} - \phi_t = v\left[\phi - s\phi_s + (2t - s^2)\phi_{ss}\right]. \tag{6.38}$$

Recall that $c_0 = \phi - s\phi_s$ (see Lemma 2.1.2). It follows that

$$[c_0]_s = -s\phi_{ss}, \qquad [c_0]_t = \phi_t - s\phi_{ts}.$$

Hence (6.38) is equivalent to

$$\left[1 - (2t - s^2)v\right][c_0]_s + s\,[c_0]_t + svc_0 = 0 \tag{6.39}$$

whenever $s \neq 0$. From now, we restrict ourself to the following solution of (6.34):

$$v = \frac{\lambda}{1+\lambda t}.$$

We put

$$\tilde{v} = \frac{1}{1+\lambda t}. \tag{6.40}$$

Then $\tilde{v}_t + v\tilde{v} = 0$. The solution c_0 of (6.39) is given by

$$c_0 = \frac{1}{1+\lambda t} f\left(\frac{2t - s^2}{(1+\lambda t)^2}\right)$$

where f is an arbitrary differential function. By the definition of c_0 and $\tilde{v}$, we have

$$\phi - s\phi_s = \tilde{v} f\left((2t - s^2)\tilde{v}^2\right). \tag{6.41}$$

Now we consider $s \in [s_0, +\infty)$ where $s_0 > 0$. Put

$$\phi = s\psi. \tag{6.42}$$

It follows that $\phi_s = \psi + s\psi_s$. Together with (6.41) yields $\tilde{v} f\left((2t - s^2)\tilde{v}^2\right) = -s^2\psi_s$. Thus

$$\psi = g(t) - \tilde{v} \int_{s_0}^{s} \sigma^{-2} f\left((2t - \sigma^2)\tilde{v}^2\right) d\sigma$$

where $g(t)$ is a differential function. Plugging this into (6.42) yields

$$\phi = sg(t) - s\tilde{v} \int_{s_0}^{s} \sigma^{-2} f\left((2t - \sigma^2)\tilde{v}^2\right) d\sigma.$$

Taking $f(\lambda) = \lambda^n$ where $n \in \{1, 2, 3, \cdots\}$ we have

$$\phi = sg(t) - s\tilde{v}^{2n+1} \int_{s_0}^{s} \sigma^{-2}(2t - \sigma^2)^n d\sigma. \tag{6.43}$$

We require the following result, the proof of which is omitted:

Lemma 6.3 *For any natural number n we have*

$$\begin{aligned} J_n :&= \int \sigma^{-2}(2t - \sigma^2)^n d\sigma \\ &= \frac{n!}{(2n-1)!!\sigma}\left[\sum_{j=2}^{n} \frac{(2n-2j-1)!!}{(n-j+1)!}(4t)^{j-1}(2t-\sigma^2)^{n-j+1} - (4t)^n\right] \\ &\quad + \frac{(2t-\sigma^2)^n}{(2n-1)\sigma} + C \end{aligned} \tag{6.44}$$

where C is a constant.

Theorem 6.2 *Let $m \geq 3$ and let $\phi(t, s)$ be a function defined by*

$$\begin{aligned} \phi(t, s) = sh(t) &+ \frac{n!}{(2n-1)!!}\frac{(4t)^n}{(1+\lambda t)^{2n+1}} - \frac{(2t-s^2)^n}{(2n-1)(1+\lambda t)^{2n+1}} \\ &- \frac{1}{(1+\lambda t)^{2n+1}} \sum_{j=2}^{n} \frac{n!(2n-2j-1)!!}{(n-j+1)!(2n-1)!!}(4t)^{j-1}(2t-s^2)^{n-j+1} \end{aligned} \tag{6.45}$$

where $n \in \{1, 2 \cdots\}$ and $h(t)$ is a differential function. Then on $\mathbb{B}^m(r_\mu)$ the following spherically symmetric Finsler metric

$$F := |y|\phi\left(\frac{|x|^2}{2}, \frac{\langle x, y\rangle}{|y|}\right)$$

is of scalar curvature. Furthermore, F is projectively flat if and only if $\lambda = 0$.

Proof By the above discussion, the functions ϕ satisfying (6.43) produce spherically symmetric Finsler metrics $F = |y|\phi\left(\frac{|x|^2}{2}, \frac{\langle x, y\rangle}{|y|}\right)$ of scalar curvature where $\tilde{\upsilon}$ is defined in (6.40). Moreover, F is projectively flat if and only if $\lambda = 0$. Plugging (6.44) into (6.43) yields

$$\begin{aligned}\phi =& sg(t) - s\tilde{\upsilon}^{2n+1}\\ &\times\left\{\frac{(2t-s^2)^n}{(2n-1)s}+\frac{n!}{s(2n-1)!!}\left[\sum_{j=2}^{n}\frac{(2n-2j-1)!!}{(n-j+1)!}(4t)^{j-1}(2t-s^2)^{n-j+1}-(4t)^n\right]\right.\\ &\left.+a(t)\right\}\\ =& sh(t)+\frac{n!}{(2n-1)!!}\tilde{\upsilon}^{2n+1}(4t)^n-\tilde{\upsilon}^{2n+1}\frac{(2t-s^2)^n}{2n-1}\\ &-\tilde{\upsilon}^{2n+1}\sum_{j=2}^{n}\frac{n!(2n-2j-1)!!}{(n-j+1)!(2n-1)!!}(4t)^{j-1}(2t-s^2)^{n-j+1},\end{aligned} \tag{6.46}$$

where $h(t)$ is a differential function. Substituting (6.40) into (6.46) yields (6.45).

We have the following two interesting special cases:

(a) When $n = 1$, then

$$F = \langle x, y\rangle h(|x|) + 8\frac{|x|^2|y|^2 - \langle x, y\rangle^2}{|y|(2+\lambda|x|^2)^3}$$

is of scalar curvature.

(b) When $n = 2$, then

$$F = \langle x, y\rangle h(|x|) + \frac{32}{(2+\lambda|x|^2)^5}\left[|y||x|^4 + 2|x|^2\frac{\langle x, y\rangle^2}{|y|} - \frac{1}{3}\frac{\langle x, y\rangle^4}{|y|^3}\right]$$

is of scalar curvature.

We known all spherically symmetric Finsler metrics are general (α, β)-metrics [27, 47, 82]. Hence Theorem 6.2 constructs a lot of new non-projectively flat general (α, β)-metrics of scalar flag curvature.

Chapter 7
Spherically Symmetric Metrics of Constant Flag Curvature

Finsler geometry is just Riemannian geometry without the quadratic restriction on its metrics [16]. One of the fundamental problems in Finlser geometry is to study and characterize Finsler metrics of constant flag curvature because Finsler metrics of constant flag curvature are the natural extension of Riemannian metrics of constant sectional curvature.

7.1 Projective Spherically Symmetric Metrics of Constant Flag Curvature

Beltrami's theorem tells us that a Riemannian metric is locally projectively flat if and only if it is of constant sectional curvature. For instance the Hilbert metric on $\mathbb{B}^n(r_{\sqrt{c}})$ given by

$$F(x, y) = \frac{\sqrt{c|y|^2 - (|x|^2|y|^2 - \langle x, y\rangle^2)}}{c - |x|^2} \tag{7.1}$$

is projectively flat on $\mathbb{B}^n(r_{\sqrt{c}})$ with constant sectional curvature -1 where $\mathbb{B}^n(r_{\sqrt{c}})$ is an n-dimensional ball with its radius $\sqrt{c}$. However the situation is much more complicated. In fact, there are lots of projectively flat Finsler metrics which are not of constant flag curvature [55]. Conversely, there are infinitely many non-projectively flat Finsler metrics with constant flag curvature [4, 32, 68]. An interesting problem then is to classify projectively flat Finsler metrics of constant flag curvature. The flag curvature in Finsler geometry is an analogue of sectional curvature in Riemannian geometry which was first introduced by L. Berwald [5].

In 2003, Z. Shen classified all locally projectively flat Randers metrics with constant flag curvature [69]. He showed that a locally projectively flat Randers metric with constant flag curvature is either locally Minkowskian or generalized

E. Guo, X. Mo, *The Geometry of Spherically Symmetric Finsler Manifolds*, SpringerBriefs in Mathematics, https://doi.org/10.1007/978-981-13-1598-5_7

Funk metric. Lately, Li-Shen classified all projectively flat (α, β)-metrics of constant flag curvature [35]. They proved that those Finsler metrics are locally Minkowskian metrics, generalized Funk metrics or Mo-Shen-Yang metrics [53]. Finsler metrics in the form $F = \alpha\phi\left(\frac{\beta}{\alpha}\right)$ are called *(α, β)-metrics* (for details, see [35, 55]). In particular, when $\phi(s) = 1 + s$, $F = \alpha + \beta$ is called a *Randers metric* [62].

In general case, Shen investigated the classification problem on projectively flat Finsler metrics of constant flag curvature and showed that such metrics can be described using algebraic equations or using Taylor expansions [17, 70]. In particular, he constructed the following projectively flat Finsler metric of constant flag curvature $K = -1$.

$$\begin{aligned} F_\varepsilon(x, y) = \frac{1}{2}&\left\{\frac{\sqrt{(1-|x|^2)|y|^2 + \langle x, y\rangle^2} + \langle x, y\rangle}{1-|x|^2}\right\} \\ -\frac{1}{2}&\left\{\frac{\varepsilon\sqrt{(1-\varepsilon^2|x|^2)|y|^2 + \varepsilon^2\langle x, y\rangle^2} + \varepsilon^2\langle x, y\rangle}{1-\varepsilon^2|x|^2}\right\}. \end{aligned} \tag{7.2}$$

Note that F_ε is no longer of (α, β) type if $\varepsilon \neq 0, -1$.

It is worth mentioning the recent result by L. Zhou [85] that any projectively flat spherically symmetric non-Riemannian Finsler metric of non-negative constant flag curvature on a convex domain $\mathcal{U} \subset \mathbb{R}^n$ is the Bryant metric (Example 1.2.5) or the metric introduced by Berwald (Example 1.2.1) [17].

In this section, we will determine all projectively flat spherically symmetric Finsler metrics of negative constant flag curvature. Precisely we prove the following theorem:

Theorem 7.1.1 *On a convex domain $\mathcal{U} \subset \mathbb{R}^n$, a spherically symmetric Finsler metric F is projectively flat with constant flag curvature $K = -1$ if and only if*

$$F = \frac{1}{2}\left[\Theta_c(x, y) - \epsilon\Theta_c(\epsilon x, y)\right], \qquad \epsilon < 1 \tag{7.3}$$

where Θ_c are the Funk metrics on the strongly convex domain $\mathbb{B}^n(r_{\sqrt{c}})$ defined by (7.6) *and* (7.9).

For the Funk metrics on the strongly convex domain $\mathbb{B}^n(r_{\sqrt{c}})$ see (7.6) and (7.9) below. We will show Theorem 7.1.1 in Sections 7.1.2 and 7.1.3. Clearly, Finsler metric (7.3) includes (7.1) and (7.2). For its explicit construction, see Sect. 7.1.1.

7.1.1 *Projective Spherically Symmetric Finsler Metrics of* $K = -1$

Below is an interesting example. This example is a slight generalization of Chern-Shen's construction [17, 70].

Example 7.1.1 Let $c \in \mathbb{R}^+$. Let ϕ_c denote a Minkowski norm defined by

$$\phi_c(y) = \frac{1}{\sqrt{c}}|y|.$$

One can construct $\mathbb{B}^n(r_{\sqrt{c}}) := \{v \in \mathbb{R}^n \mid \phi_c(v) < 1\}$. Then $\mathbb{B}^n(r_{\sqrt{c}})$ is a strongly convex domain [71]. Thus $\left(\mathbb{B}^n(r_{\sqrt{c}}),\ \Phi_c(x,\ y)\right)$ is a Minkowski manifold where

$$\Phi_c(x,\ y) = \phi_c(y).$$

For each $x \in \mathbb{B}^n(r_{\sqrt{c}})$, identify $T_x\mathbb{B}^n(r_{\sqrt{c}})$ with $\mathbb{R}^n$. Thus $V_x := -x$ is a radical vector field on $\mathbb{B}^n(r_{\sqrt{c}})$ satisfying $\Phi_c(x,\ -V_x) = \phi_c(x) < 1$. Moreover V is a homothetic field of Φ_c with dilation $-\frac{1}{2}$ [47, 51]. By using Φ_c and V, we produce the Funk metric Θ_c on the strongly convex domain $\mathbb{B}^n(r_{\sqrt{c}})$ in terms of the following navigation problem

$$\Phi_c\left(x,\ \frac{y}{\Theta_c(x,\ y)} - V_x\right) = 1.$$

In particular, Θ_1 is the well-known Funk metric on the unit ball.

Note that $\Phi_c(x,\ y)$ is Riemannian, so that Θ_c is of Randers type. Then

$$\Theta_c = \alpha + \beta \tag{7.4}$$

where $\alpha = \sqrt{a_{ij}y^i y^j}$ and $\beta = b_i y^i$. By using (1.40) and (1.41) in [17], we get

$$a_{ij} = \frac{(c - |x|^2)\delta_{ij} + x^i x^j}{(c - |x|^2)^2}, \qquad b_i = \frac{x^i}{c - |x|^2}. \tag{7.5}$$

Plugging (7.5) into (7.4) yields

$$\Theta_c = \frac{\sqrt{c|y|^2 - (|x|^2|y|^2 - \langle x,\ y\rangle^2)} + \langle x,\ y\rangle}{c - |x|^2}. \tag{7.6}$$

By Theorem 8.2.3 in [17], we conclude that the following function

$$F = \frac{1}{2}\left[\Theta_c(x,\ y) - \varrho\Theta_c(\varrho\varepsilon x,\ \varepsilon y)\right] \tag{7.7}$$

is a projectively flat Finsler metric on its domain with constant flag curvature $K = -1$ where $\varepsilon = \pm 1$ and ϱ is a constant chosen so that

$$\psi := \frac{\phi_c(y) - \varrho\phi_c(\varepsilon y)}{2}$$

is a Minkowski norm on $\mathbb{R}^n$. Substituting (7.6) into (7.7) yields

$$\begin{aligned} F(x, y) &= \frac{1}{2}\left\{\frac{\sqrt{c|y|^2 - (|x|^2|y|^2 - \langle x, y\rangle^2)} + \langle x, y\rangle}{c - |x|^2}\right\} \\ &\quad - \frac{\varrho}{2}\left\{\frac{\sqrt{c|\varepsilon y|^2 - (|\varrho\varepsilon x|^2|\varepsilon y|^2 - \langle \varrho\varepsilon x, \varepsilon y\rangle^2)} + \langle \varrho\varepsilon x, \varepsilon y\rangle}{c - |\varrho\varepsilon x|^2}\right\} \\ &= \frac{1}{2}\left\{\frac{\sqrt{c|y|^2 - (|x|^2|y|^2 - \langle x, y\rangle^2)} + \langle x, y\rangle}{c - |x|^2}\right\} \\ &\quad - \frac{1}{2}\left\{\frac{\varrho\sqrt{c|y|^2 - \varrho^2(|x|^2|y|^2 - \langle x, y\rangle^2)} + \varrho^2\langle x, y\rangle}{c - \varrho^2|x|^2}\right\} \\ &= \frac{1}{2}\left[\Theta_c(x, y) - \varrho\Theta_c(\varrho x, y)\right]. \end{aligned} \tag{7.8}$$

A direct calculation yields that F is a projectively flat Finsler metric of negative constant flag curvature $K = -1$ when $\varrho < 1$ (see [71], Example 2.5).

Similarly we can construct projectively flat spherically symmetric Finsler metrics of $K = -1$ by

$$F(x, y) = \frac{1}{2}\left[\Theta_c(x, y) - \epsilon\Theta_c(\epsilon x, y)\right],$$

where Θ_c is the following Funk metric on $\mathbb{B}^n(r_{\sqrt{c}})$ [10, 69]

$$\Theta_c = \frac{\sqrt{c|y|^2 - (|x|^2|y|^2 - \langle x, y\rangle^2)} - \langle x, y\rangle}{c - |x|^2}. \tag{7.9}$$

Let $F = |y|\phi(z_1, z_2)$ be a projectively flat spherically symmetric Finsler metric, where

$$z_1 = \frac{1}{|y|}\sqrt{|x|^2|y|^2 - \langle x, y\rangle^2}, \qquad z_2 = \frac{\langle x, y\rangle}{|y|}. \tag{7.10}$$

For more details, see [85]. The reader should note that our notation ϕ is precisely Zhou's notation $\tilde{\phi}$. From (7.10) we have

$$z_1^2 + z_2^2 = |x|^2. \tag{7.11}$$

The following lemma will be used in Sect. 7.1.3.

Lemma 7.1.1 *For $a, b \geq 0$, we have*

$$\sqrt{a} \pm \sqrt{b} = sgn(a \pm b)\sqrt{a + b \pm 2\sqrt{ab}}. \tag{7.12}$$

Proof

$$(\sqrt{a} \pm \sqrt{b})^2 = a \pm 2\sqrt{a}\sqrt{b} + b = a + b \pm 2\sqrt{ab}.$$

In order to determine projectively flat spherically symmetric Finsler metrics with negative constant flag curvature, we consider the following ordinary differential equation (see (7.17) below)

$$2yy'' - 3(y')^2 - 4y^4 = 0. \tag{7.13}$$

Lemma 7.1.2 *The non-trivial solution of* (7.13) *is*

$$y = \frac{1}{c_1x^2 + c_2x + c_3}, \tag{7.14}$$

where $c_j \in \mathbb{R}$, $j = 1, 2, 3$ and satisfy

$$c_2^2 - 4c_1c_3 - 4 = 0. \tag{7.15}$$

Proof See Lemma 2.2 in [85].

L. Zhou obtained the following necessary condition on ϕ for the spherically symmetric Finsler metric $F = |y|\phi(z_1, z_2)$ to be projectively flat with constant flag curvature $K = -1$ (see [85], (6) and (8‴)).

Lemma 7.1.3 *Let $F = |y|\phi(z_1, z_2)$ be a projectively flat spherically symmetric Finsler metric with constant flag curvature -1. Then*

$$\begin{aligned} \text{(i) } \phi_{z_1}(\phi_{z_2}^2 z_2^2 + 4\phi_{z_2}\phi z_2 + 4\phi^2 - 4\phi^4 z_2^2 \\ = z_1(\phi_{z_2}^3 z_2 + 4\phi^5 + 3\phi_{z_2}^2\phi - 4\phi^4\phi_{z_2}z_2), \end{aligned} \tag{7.16}$$

$$\text{(ii) } 3\phi_{z_2}^2 - 2\phi\phi_{z_2z_2} + 4\phi^4 = 0. \tag{7.17}$$

Now our classification is reduced to finding all of the solutions of (7.17) and (7.16) such that

$$F = |y|\phi(z_1, z_2), \tag{7.18}$$

satisfies the conditions of Finsler metrics. By Lemma 7.1.2, the solution of (7.17) is

$$\phi(z_1, z_2) = \frac{1}{\tilde{c}_1(z_1)z_2^2 + \tilde{c}_2(z_1)z_2 + \tilde{c}_3(z_1)}, \tag{7.19}$$

where $\tilde{c}_1(z_1)$, $\tilde{c}_2(z_1)$ and $\tilde{c}_3(z_1)$ satisfy

$$\tilde{c}_2^2(z_1) - 4\tilde{c}_1(z_1)\tilde{c}_3(z_1) - 4 = 0. \tag{7.20}$$

7.1.2 ϕ Satisfies $\tilde{c}_1(z_1) = 0$

Our proof of Theorem 7.1.1 is given in two steps. In this subsection, we shall classify the case when ϕ satisfies $\tilde{c}_1(z_1) = 0$. In the following subsection, we are going to study the case when $\tilde{c}_1(z_1) \neq 0$.

Proposition 7.1.1 *Let F be a projectively flat spherically symmetric Finsler metric with $K = -1$ on a convex domain $\mathcal{U} \subset \mathbb{R}^n$. Assume that*

$$\tilde{c}_1(z_1) = 0. \tag{7.21}$$

Then

$$F = \frac{\sqrt{c|y|^2 - (|x|^2|y|^2 - \langle x, y\rangle^2)} \pm \langle x, y\rangle}{2(c - |x|^2)}.$$

Proof Plugging (7.21) into (7.20) yields

$$\tilde{c}_2(z_1) = \pm 2. \tag{7.22}$$

Substituting (7.21)and (7.22) into (7.19), we find that

$$\phi = \frac{1}{\pm 2z_2 + \tilde{c}_3(z_1)}. \tag{7.23}$$

If $\phi = \frac{1}{2z_2 + \tilde{c}_3(z_1)}$, we have

$$\phi = \frac{1}{2[z_2 + c_1(z_1)]}, \tag{7.24}$$

where $c_1(z_1) = \frac{1}{2}\tilde{c}_3(z_1)$. By (7.24), we get

$$\phi_{z_1} = -2c_1'(z_1)\phi^2 = c_1'(z_1)\phi_{z_2} \tag{7.25}$$

and

$$\phi_{z_2} = -2\phi^2. \tag{7.26}$$

Plugging (7.25) and (7.26) into (7.16), we obtain

$$2z_2 c_1'(z_1)\phi - c_1'(z_1) = 2z_1\phi. \tag{7.27}$$

Substituting (7.24) into (7.27) yields

$$c_1(z_1)c_1'(z_1) + z_1 = 0. \tag{7.28}$$

The solutions of (7.28) are

$$c_1(z_1) = \pm\sqrt{c - z_1^2}, \tag{7.29}$$

where c is a positive constant. From which together with (7.18) and (7.24) we obtain

$$\begin{aligned} F &= \frac{1}{2}\frac{|y|}{z_2 + c_1(z_1)} \\ &= \frac{1}{2}\frac{|y|}{z_2 \pm \sqrt{c - z_1^2}} \\ &= \frac{|y|}{2}\frac{z_2 \mp \sqrt{c - z_1^2}}{z_2^2 - (c - z_1^2)} = \frac{|y|}{2}\frac{\pm\sqrt{c - z_1^2} - z_2}{c - (z_1^2 + z_2^2)}. \end{aligned} \tag{7.30}$$

Plugging (7.10) and (7.11) into (7.30) yields

$$F = F_{\pm} = \frac{\pm\sqrt{c|y|^2 - (|x|^2|y|^2 - \langle x, y\rangle^2)} - \langle x, y\rangle}{2(c - |x|^2)}.$$

Hence F_+ is of Randers type. In particular, $2F_+$ is the Funk metric on $\mathbb{B}^n(r_{\sqrt{c}})$ (see (7.6) and (7.9)). However, F_- is not a Finsler metric.

If $\phi = \frac{1}{-2z_2 + \tilde{c}_3(z_1)}$, we get that

$$\phi = \frac{1}{2[-z_2 + c_1(z_1)]}, \tag{7.31}$$

where $c_1(z_1) = \frac{1}{2}\tilde{c}_3(z_1)$. Then from (7.31), we have

$$\phi_{z_1} = -2c_1'(z_1)\phi^2 = -c_1'(z_1)\phi_{z_2} \tag{7.32}$$

and

$$\phi_{z_2} = 2\phi^2. \tag{7.33}$$

Plugging (7.32) and (7.33) into (7.16) yields

$$-c_1'(z_1)(2\phi z_2 + 1) = 2\phi z_1. \tag{7.34}$$

Substituting (7.31) into (7.34), we get (7.28). Thus it has solutions (7.29). Together with (7.31) we have

$$F = \frac{|y|}{2} \frac{\pm\sqrt{c - z_1^2} + z_2}{c - (z_1^2 + z_2^2)}. \tag{7.35}$$

Substituting (7.10) and (7.11) into (7.35), we can obtain

$$F = F_\pm = \frac{\pm\sqrt{c|y|^2 - (|x|^2|y|^2 - \langle x, y\rangle^2)} + \langle x, y\rangle}{2(c - |x|^2)}.$$

Again, F_+ is of Randers type, and $2F_+$ is the Funk metric on $\mathbb{B}^n(r_{\sqrt{c}})$ (cf. [10, 69], Example 3.4.3 in [17]). However, F_- is not a Finsler metric.

7.1.3 ϕ Satisfies $\tilde{c}_1(z_1) \neq 0$

In this subsection, we are going to classify the case when $\tilde{c}_1(z_1) \neq 0$. In this case we shall show the following:

Proposition 7.1.2 *Let F be a projective spherically symmetric Finsler metric with $K = -1$ on a convex domain $\mathcal{U} \subset \mathbb{R}^n$. Assume that*

$$\tilde{c}_1(z_1) \neq 0. \tag{7.36}$$

Then $F = \frac{1}{2}[\Theta_c(x, y) - \varrho\Theta_c(\epsilon x, y)]$ where $\varrho \neq 0$ and Θ_c is the Funk metric on the strongly convex domain $\mathbb{B}^n(r_{\sqrt{c}})$ (see (7.6) and (7.9)).

Proof From (7.20) and (7.36), we have

$$\frac{\tilde{c}_3}{\tilde{c}_1} = \frac{\tilde{c}_2^2}{4\tilde{c}_1^2} - \frac{1}{\tilde{c}_1^2}$$

where $\tilde{c}_j = \tilde{c}_j(z_1),\ j = 1, 2, 3$. Together with (7.9) we get

$$\phi = \frac{1}{\tilde{c}_1} \times \frac{1}{z_2^2 + \frac{\tilde{c}_2}{\tilde{c}_1} z_2 + \frac{\tilde{c}_3}{\tilde{c}_1}} = \frac{1}{\tilde{c}_1} \times \frac{1}{z_2^2 + \frac{\tilde{c}_2}{\tilde{c}_1} z_2 + \frac{\tilde{c}_2^2}{4\tilde{c}_1^2} - \frac{1}{\tilde{c}_1^2}} = \frac{1}{\tilde{c}_1} \times \frac{1}{(z_2 + c_2)^2 - \left(\frac{1}{\tilde{c}_1}\right)^2},$$

where $c_2 := \dfrac{\tilde{c}_2}{2\tilde{c}_1}$.

Case 1: $\tilde{c}_1(z_1) < 0$.
Note that $\phi > 0$, we have $[z_2 + c_2(z_1)]^2 < \frac{1}{\tilde{c}_1(z_1)^2}$ and

$$\phi = \frac{c_1}{c_1^2 - (z_2 + c_2)^2}, \tag{7.37}$$

where

$$c_1 := -\frac{1}{\tilde{c}_1(z_1)} > 0. \tag{7.38}$$

That is,

$$\frac{1}{c_1^2 - (z_2 + c_2)^2} = \frac{\phi}{c_1}. \tag{7.39}$$

By using (7.37) and (7.39) we obtain

$$\phi_{z_1} = \frac{c_1'}{c_1}\phi - 2\frac{c_1 c_1' - (z_2 + c_2)c_2'}{c_1}\phi^2 \tag{7.40}$$

and

$$\phi_{z_2} = \frac{2(z_2 + c_2)}{c_1}\phi^2, \tag{7.41}$$

where $c_j' = c_j'(z_1),\ j = 1, 2$. We rewrite (7.16) by

$$\phi_{z_1}(\mathrm{I}) = z_1(\mathrm{II}), \tag{7.42}$$

where

$$\begin{aligned}
(I) :&= \phi_{z_2}^2 z_2^2 + 4\phi_{z_2}\phi z_2 + 4\phi^2 - 4\phi^4 z_2^2 \\
&= \frac{4(z_2 + c_2)^2}{c_1^2}\phi^4 z_2^2 + 8\frac{z_2 + c_2}{c_1}\phi^3 z_2 + 4\phi^2 - 4\phi^4 z_2^2 \\
&= 4\phi^2 \left[\frac{(z_2 + c_2)^2 - c_1^2}{c_1^2}\phi^2 z_2^2 + 2\frac{z_2 + c_2}{c_1}\phi z_2 + 1\right] \\
&= 4\phi^2 \left(1 - \frac{1}{c_1}\phi z_2^2 + 2\frac{z_2 + c_2}{c_1}\phi z_2\right) = 4\phi^2 \left(1 + \frac{z_2 + 2c_2}{c_1}\phi z_2\right)
\end{aligned} \tag{7.43}$$

and

$$
\begin{aligned}
(II) :&= \phi_{z_2}^3 z_2 + 4\phi^5 + 3\phi_{z_2}^2\phi - 4\phi^4\phi_{z_2}z_2 \\
&= 4\phi^5\left[\frac{2(z_2+c_2)^3}{c_1^3}z_2\phi + 1 + 3\frac{(z_2+c_2)^2}{c_1^2} - 2\frac{z_2+c_2}{c_1}z_2\phi\right] \\
&= 4\phi^5\left[\frac{2(z_2+c_2)}{c_1}z_2\phi\frac{(z_2+c_2)^2-c_1^2}{c_1^2} + 1 + 3\frac{(z_2+c_2)^2}{c_1^2}\right] \\
&= 4\phi^5\left[\frac{2(z_2+c_2)}{c_1}z_2\frac{-1}{c_1} + 1 + 3\frac{(z_2+c_2)^2}{c_1^2}\right] \\
&= \frac{4\phi^5}{c_1^2}[c_1^2 + (z_2+c_2)(z_2+3c_2)].
\end{aligned}
\tag{7.44}
$$

Plugging (7.40), (7.41), (7.43) and (7.44) into (7.42) and then multiplying $\frac{1}{4\phi^3}$, we have

$$
\begin{aligned}
&(1 + \frac{z_2+2c_2}{c_1}\phi z_2)\left[\frac{c_1'}{c_1} - 2\frac{c_1c_1' - (z_2+c_2)c_2'}{c_1}\phi\right] \\
&\qquad = \frac{z_1}{c_1^2}\phi^2[c_1^2 + (z_2+c_2)(z_2+3c_2)].
\end{aligned}
\tag{7.45}
$$

Substituting (7.37) into (7.45) yields

$$
\begin{aligned}
&\left[1 + \frac{z_2+2c_2}{c_1^2-(z_2+c_2)^2}z_2\right]\left[\frac{c_1'}{c_1} - 2\frac{c_1c_1' - (z_2+c_2)c_2'}{c_1^2-(z_2+c_2)^2}\right] \\
&\qquad = z_1\frac{c_1^2 + (z_2+c_2)(z_2+3c_2)}{[c_1^2-(z_2+c_2)^2]^2}.
\end{aligned}
\tag{7.46}
$$

It follows that, for arbitrary z_2

$$
f(z_1)z_2^2 + g(z_1)z_2 + h(z_1) = 0, \tag{7.47}
$$

where

$$
\begin{aligned}
f(z_1) &= c_1'(c_2^2 - c_1^2) - c_1z_1, \\
g(z_1) &= 2[(c_1'c_2 - c_1c_2')(c_2^2 - c_1^2) - 2c_1c_2z_1], \\
h(z_1) &= c_1'(c_2^4 - c_1^4) - 2c_1c_2c_2'(c_2^2 - c_1^2) - c_1^3z_1 - 3c_1c_2^2z_1.
\end{aligned}
$$

Hence we have $f(z_1) = g(z_1) = h(z_1) = 0$, that is,

$$c_1'(c_2^2 - c_1^2) = c_1 z_1, \tag{7.48}$$

$$(c_1' c_2 - c_1 c_2')(c_2^2 - c_1^2) = 2c_1 c_2 z_1, \tag{7.49}$$

$$2c_1 c_2 c_2'(c_2^2 - c_1^2) + c_1^3 z_1 + 3c_1 c_2^2 z_1 = c_1'(c_2^4 - c_1^4). \tag{7.50}$$

By using (7.38) and (7.48), we have

$$c_1(c_2^2 - c_1^2) \neq 0. \tag{7.51}$$

Combining this with (7.48) we have

$$c_1' = \frac{c_1 z_1}{c_2^2 - c_1^2}. \tag{7.52}$$

Plugging this into (7.49), and then multiplying $\frac{1}{c_1}$, we get

$$\left(\frac{c_2 z_1}{c_2^2 - c_1^2} - c_2'\right)(c_2^2 - c_1^2) = 2c_2 z_1. \tag{7.53}$$

From which together with (7.51) we obtain

$$c_2' = \frac{-c_2 z_1}{c_2^2 - c_1^2}. \tag{7.54}$$

Substituting (7.52) and (7.54) into (7.50), we obtain an identical relation. Hence (7.50) is not independent. Using (7.52) and (7.54) we get

$$(c_1 c_2)' = c_1' c_2 + c_1 c_2' = \frac{c_1 z_1}{c_2^2 - c_1^2} c_2 + c_1 \frac{-c_2 z_1}{c_2^2 - c_1^2} = 0.$$

It follows that $d_1 := c_1 c_2 = \text{constant}$. Together with (7.51) we obtain

$$c_2 = \frac{d_1}{c_1}. \tag{7.55}$$

Plugging this into (7.48), and then multiplying c_1^{-1} we have

$$(d_1^2 \frac{1}{c_1^3} - c_1)c_1' = z_1. \tag{7.56}$$

Integrating (7.56) yields

$$-\frac{d_1^2}{2c_1^2} - \frac{1}{2}c_1^2 = \frac{1}{2}z_1^2 - d_2,$$

where d_2 is a non-negative constant. It follows that

$$\xi^2 + (z_1^2 - 2d_2)\xi + d_1^2 = 0, \tag{7.57}$$

where

$$\xi := c_1^2. \tag{7.58}$$

The solutions of (7.57) are

$$c_1^2 = \xi = \frac{2d_2 - z_1^2 \pm \sqrt{(2d_2 - z_1^2)^2 - 4d_1^2}}{2}.$$

Together with (7.38) we have

$$c_1(z_1) = \frac{1}{\sqrt{2}}\sqrt{2d_2 - z_1^2 \pm \sqrt{(2d_2 - z_1^2)^2 - 4d_1^2}}. \tag{7.59}$$

Combining this with (7.55) we obtain

$$c_2(z_1) = \frac{\delta}{\sqrt{2}}\sqrt{2d_2 - z_1^2 \mp \sqrt{(2d_2 - z_1^2)^2 - 4d_1^2}}, \tag{7.60}$$

where

$$\delta = \operatorname{sgn} d_1. \tag{7.61}$$

If $\delta = 1$, from (7.59), (7.60) and Lemma 7.1.4, we get

$$\begin{aligned}(c_1 + c_2)(z_1) &= \frac{1}{\sqrt{2}}\sqrt{2d_2 - z_1^2 \pm \sqrt{(2d_2 - z_1^2)^2 - 4d_1^2}} \\ &\quad + \frac{1}{\sqrt{2}}\sqrt{2d_2 - z_1^2 \mp \sqrt{(2d_2 - z_1^2)^2 - 4d_1^2}} \\ &= \frac{1}{\sqrt{2}}\sqrt{2(2d_2 - z_1^2) + 2\sqrt{4d_1^2}} = \sqrt{2(d_2 + d_1) - z_1^2}.\end{aligned} \tag{7.62}$$

Similarly, we have

$$(c_1 - c_2)(z_1) = \pm\sqrt{2(d_2 - d_1) - z_1^2}, \tag{7.63}$$

where $d_2 > d_1$. Together with (7.37), (7.62) and (7.63) we get

$$2\phi = \frac{2c_1}{c_1^2 - (z_2 + c_2)^2}$$
$$= \frac{\sqrt{2(d_2 + d_1) - z_1^2} \pm \sqrt{2(d_2 - d_1) - z_1^2}}{[\sqrt{2(d_2 + d_1) - z_1^2} + z_2][\pm\sqrt{2(d_2 - d_1) - z_1^2} - z_2]} \tag{7.64}$$
$$= \frac{1}{\sqrt{2(d_2 + d_1) - z_1^2} + z_2} + \frac{1}{\pm\sqrt{2(d_2 - d_1) - z_1^2} - z_2} =: 2\phi_\pm.$$

It is easy to see that $F_- = |y|\phi_-$ is not a Finsler metric. However

$$2\phi_+ = \frac{\sqrt{2(d_2 + d_1) - z_1^2} - z_2}{2(d_2 + d_1) - (z_1^2 + z_2^2)} + \frac{\sqrt{2(d_2 - d_1) - z_1^2} + z_2}{2(d_2 - d_1) - (z_1^2 + z_2^2)}.$$

It follows that

$$F = F_+$$
$$:= |y|\phi_+$$
$$= \frac{\sqrt{c|y|^2 - (|x|^2|y|^2 - \langle x, y\rangle^2)} \pm \langle x, y\rangle}{2(c - |x|^2)} \tag{7.65}$$
$$- \frac{\epsilon\sqrt{c|y|^2 - \epsilon^2(|x|^2|y|^2 - \langle x, y\rangle^2)} \pm \epsilon^2\langle x, y\rangle}{2(c - \epsilon^2|x|^2)},$$

where $c := 2(d_2 \mp d_1)$, $\epsilon = -\sqrt{\frac{d_2 \mp d_1}{d_2 \pm d_1}}$, and we have used (7.10) and (7.11).

If $\delta = 0$, we get that $d_1 = 0$ and $c_2(z_1) = 0$. Plugging this into (7.59) yields

$$c_1(z_1) = \sqrt{2d_2 - z_1^2}.$$

where $d_2 > 0$. Hence we have

$$F = |y|\phi = \frac{|y|c_1(z_1)}{c_1^2(z_1) - z_2^2}$$
$$= \frac{|y|\sqrt{2d_2 - z_1^2}}{2d_2 - (z_1^2 + z_2^2)} = \frac{\sqrt{c|y|^2 - (|x|^2|y|^2 - \langle x, y\rangle^2)}}{c - |x|^2} \tag{7.66}$$

where $c = 2d_2$. We obtain the Hilbert metric on $\mathbb{B}^n(r_{\sqrt{c}})$.

If $\delta = -1$, from (7.59), (7.60) and Lemma 7.1.1, we get

$$(c_1 + c_2)(z_1) = \pm\sqrt{2(d_2 + d_1) - z_1^2}, \; (c_1 - c_2)(z_1) = \sqrt{2(d_2 - d_1) - z_1^2}, \tag{7.67}$$

$$2c_1(z_1) = \sqrt{2(d_2 - d_1) - z_1^2} \pm \sqrt{2(d_2 + d_1) - z_1^2}, \tag{7.68}$$

where $d_2 + d_1 > 0$. By using (7.10), (7.11), (7.67) and (7.68), we obtain

$$\begin{aligned} F_{+}(x, y) = &\frac{\sqrt{c|y|^2 - (|x|^2|y|^2 - \langle x, y\rangle^2)} \pm \langle x, y\rangle}{2(c - |x|^2)} \\ &- \frac{\varrho\sqrt{c|y|^2 - \varrho^2(|x|^2|y|^2 - \langle x, y\rangle^2)} \pm \varrho^2\langle x, y\rangle}{2(c - \varrho^2|x|^2)}, \end{aligned} \tag{7.69}$$

where $c := 2(d_2 \mp d_1)$, $\varrho = -\sqrt{\frac{d_2 \mp d_1}{d_2 \pm d_1}}$, however, F_{-} is not a Finsler metric.

Case 2: $\tilde{c}_1(z_1) > 0$.

Since $\phi > 0$, we conclude that $[z_2 + c_2(z_1)]^2 > \frac{1}{\tilde{c}_1^2(z_1)}$ and

$$\phi = \frac{c_1}{(z_2 + c_2)^2 - c_1^2}, \tag{7.70}$$

where $c_1(z_1) := \frac{1}{\tilde{c}_1(z_1)}$. By direct calculations, we have (7.59), (7.60) and (7.61).

If $\delta = 1$, then we get (7.62) and (7.63). Together with (7.70) we have

$$\begin{aligned} 2\phi &= 2\phi_{\pm} \\ &= \frac{1}{\mp\sqrt{2(d_2 - d_1) - z_1^2} + z_2} - \frac{1}{\sqrt{2(d_2 + d_1) - z_1^2} + z_2}. \end{aligned} \tag{7.71}$$

By using (7.71), (7.10) and (7.11), we get

$$\begin{aligned} F_{+} :&= |y|\phi_{+} \\ &= \frac{\sqrt{c|y|^2 - (|x|^2|y|^2 - \langle x, y\rangle^2)} - \langle x, y\rangle}{2(c - |x|^2)} \\ &\quad - \frac{\varrho\sqrt{c|y|^2 - \varrho^2(|x|^2|y|^2 - \langle x, y\rangle^2)} - \epsilon^2\langle x, y\rangle}{2(c - \varrho^2|x|^2)}, \end{aligned}$$

where $c := 2(d_2 - d_1)$, $\varrho = \sqrt{\frac{d_2 - d_1}{d_2 + d_1}}$.

If $\delta = 0$, we get $d_1 = 0,\ \ c_2(z_1) = 0,\ \ c_1(z_1) = \sqrt{2d_2 - z_1^2}$. It follows that F is not a Finsler metric.

If $\delta = -1$, then we get (7.67) and (7.68). Together with (7.10), (7.70) and (7.11), we get

$$\begin{aligned} F_+ :&= |y|\phi_+ \\ &= \frac{\sqrt{c|y|^2 - (|x|^2|y|^2 - \langle x, y\rangle^2)} + \langle x, y\rangle}{2(c - |x|^2)} \\ &\quad - \frac{\varrho\sqrt{c|y|^2 - \varrho^2(|x|^2|y|^2 - \langle x, y\rangle^2)} + \varrho^2\langle x, y\rangle}{2(c - \varrho^2|x|^2)}, \end{aligned}$$

where $c := 2(d_2 + d_1),\ \ \varrho = \sqrt{\dfrac{d_2 + d_1}{d_2 - d_1}} < 1.$

7.2 Some Explicit Constructions of Spherically Symmetric Metrics of Constant Curvature

The classification theorem of projective spherically symmetric metrics of constant flag curvature has been completed by L.Zhou and Mo-Zhu [58, 85] (see Sect. 7.1). By finding two partial differential equations equivalent to spherically symmetric metrics being of constant flag curvature, we are going to construct explicitly new spherically symmetric metrics of constant flag curvature in this section. Moreover, these Finsler metrics are locally projectively flat.

7.2.1 *Preliminaries*

Let F be a Finsler metric on an m-dimensional manifold M. The *Riemann curvature* of F is a family of endomorphism $R_y = R^i_j dx^j \otimes \frac{\partial}{\partial x^i} : T_xM \to T_xM$, given in (6.12) where G^i are the geodesic coefficients of F. A Finsler metric F is of scalar (flag) curvature with flag curvature K is equivalent to the following identity (see (6.27))

$$R^i_j = K(F^2\delta^i_j - FF_{y^j}y^i) \tag{7.72}$$

where $K = K(x,\ y)$ is a scalar function on the tangent bundle.

Let us recall a formula for the Riemann curvature of a spherically symmetric Finsler metric $F = |y|\phi\left(|x|,\ \frac{\langle x, y\rangle}{|y|}\right)$.

Let

$$R_1 := P^2 - \frac{1}{r}(sP_r + rP_s) + 2Q[1 + sP + (r^2 - s^2)P_s], \tag{7.73}$$

$$R_2 := 2Q(2Q - sQ_s) + \frac{1}{r}(2Q_r - sQ_{rs} - rQ_{ss}) + (r^2 - s^2)(2QQ_{ss} - Q_s^2), \tag{7.74}$$

$$R_3 := -sR_2, \tag{7.75}$$

$$\begin{aligned} R_4 := &\frac{2}{r}P_r - Q_s - P_{ss} - \frac{s}{r}P_{rs} + 2Q(P - sP_s) + 2(r^2 - s^2)QP_{ss} \\ &- sPQ_s - (r^2 - s^2)P_sQ_s - PP_s, \end{aligned} \tag{7.76}$$

$$R_5 := -R_1 - sR_4, \tag{7.77}$$

where $P_s := \frac{\partial P}{\partial s}$, $P_r := \frac{\partial P}{\partial r}$, $Q_s := \frac{\partial Q}{\partial s}$, $Q_r := \frac{\partial Q}{\partial r}$, $Q_{ss} := \frac{\partial^2 Q}{\partial s^2}$, P and Q are given in (3.3) and (3.2) respectively. We have the following [28, 38]

Lemma 7.2.1 *Let $F = |y|\phi\left(|x|, \frac{\langle x, y\rangle}{|y|}\right)$ be a spherically symmetric Finsler metric on $\mathbb{B}^m(r_\mu)$. Then the Riemann curvature of F is given by*

$$R^i_j = u^2 R_1 \delta^{ij} + u^2 R_2 x^i x^j + uR_3 x^i y^j + uR_4 x^j y^i + R_5 y^i y^j \tag{7.78}$$

where $u = |y|$.

7.2.2 Ξ-Curvature

Let $F = F(x, y)$ be a Finsler metric on a manifold M. Let $\mathbf{S}$ be the S-curvature of F [12, 13, 49] (see Sect. 3.1). We consider the following non-Riemannian quantity, $\Xi = \Xi_j dx^j$, on the tangent bundle TM:

$$\Xi_j := \mathbf{S}_{\cdot j|i} y^i - \mathbf{S}_{|j}$$

where "." denotes the vertical covariant derivative and "|" denotes the horizonal covariant derivative. Ξ is called the *Ξ-curvature of F* [73] (*χ-curvature* in an alternative terminology in [13]).

The Ξ-curvature gives a measure of failure of a Finsler metric of scalar curvature to be of isotropic flag curvature and delicately related to Riemannian quantities [49, 73, 78, 80]. We have the following:

Lemma 7.2.2 *Let (M, F) be a Finsler manifold of scalar flag curvature with flag curvature K. Then F has isotropic flag curvature if and only if the Ξ-curvature vanishes.*

Proof A straightforward computation shows the following ([73], Corollary 2.4)

$$\Xi_j = -\frac{m+1}{3} F^2 K_{.j}.$$

Since $K_{.j} = \frac{\partial K}{\partial y^j}$, we get that the Ξ-curvature vanishes if and only if the flag curvature K is a function of $x \in M$ only.

A Finsler metric F is said to have *isotropic flag curvature* if its flag curvature $K(P, y) = K(x)$ is a scalar function on M.

Now we compute the Ξ-curvature of an m-dimensional orthogonally invariant Finsler metric $F(x, y) = |y|\phi\left(|x|, \frac{\langle x, y\rangle}{|y|}\right)$.

By (7.75), (7.77) and Lemma 7.2.1, we can easily get a formula for the Ricci curvature $Ric = \sum_{j=1}^{m} R^j_j$.

$$Ric = mu^2 R_1 + u^2|x|^2 R_2 + u\langle x, y\rangle R_3 + u\langle x, y\rangle R_4 + |y|^2 R_5 = u^2 R \tag{7.79}$$

where

$$R := (m-1)R_1 + (r^2 - s^2)R_2. \tag{7.80}$$

We have

$$\frac{\partial}{\partial y^j} Ric = \frac{\partial}{\partial y^j}(u^2 R) = \frac{\partial u^2}{\partial y^j} R + u^2 \frac{\partial R}{\partial s} s_{y^j} = u R_s x^j + (2R - sR_s) y^j \tag{7.81}$$

where $R_s := \frac{\partial R}{\partial s}$ and we have used (3.10) and (3.13). By simple calculations, we have

$$s_{y^k} y^k = 0, \quad s_{y^k} x^k = \frac{r^2 - s^2}{u}. \tag{7.82}$$

We denote $\frac{\partial R_j}{\partial s}$ by R_{js} $j = 1, \cdots, 5$. By using (7.78), we obtain

$$\begin{aligned}
\frac{\partial R^i_j}{\partial y^k} &= 2y^k R_1 \delta^i_j + u^2 R_{1s} s_{y^k} \delta^i_j + 2y^k R_2 x^i x^j + u^2 R_{2s} s_{y^k} x^i x^j + \frac{y^k}{u} R_3 x^i y^j \\
&\quad + u R_{3s} s_{y^k} x^i y^j + u R_3 x^i \delta^j_k + \frac{y^k}{u} R_4 x^j y^i + u R_{4s} s_{y^k} x^j y^i + u R_4 x^j \delta^i_k \\
&\quad + R_{5s} s_{y^k} y^i y^j + R_5 \delta^i_k y^j + R_5 y^i \delta^j_k.
\end{aligned}$$

It follows that

$$\sum_i \frac{\partial R^i_j}{\partial y^i} = u[R_{1s} + 2sR_2 + (r^2 - s^2)R_{2s} + R_3 + (m+1)R_4]x^j \\ + [2R_1 - sR_{1s} + sR_3 + (r^2 - s^2)R_{3s} + (m+1)R_5]y^j \tag{7.83}$$

where we have used (7.82) and (3.13). By (7.75), we have $R_{3s} = -R_2 - sR_{2s}$. Taking this together with (7.75), (7.77) and (7.83), we obtain

$$\sum_i \frac{\partial R^i_j}{\partial y^i} = u\mathfrak{M}x^j + \mathfrak{N}y^j, \tag{7.84}$$

where

$$\mathfrak{M} := R_{1s} + sR_2 + (r^2 - s^2)R_{2s} + (m+1)R_4, \tag{7.85}$$

and

$$\mathfrak{N} := (1-m)R_1 - sR_{1s} - r^2R_2 - s(r^2 - s^2)R_{2s} - (m+1)sR_4. \tag{7.86}$$

Below is a delicate relationship between Ξ-curvature and Riemann curvature:

Lemma 7.2.3 ([48, 49, 52, 73])

$$\Xi_j = -\frac{1}{3}\left(2\sum_i \frac{\partial R^i_j}{\partial y^i} + \frac{\partial}{\partial y^j} Ric\right). \tag{7.87}$$

Plugging (7.81) and (7.84) into (7.87), we obtain

$$\Xi_j = -\frac{1}{3}\left[u(2\mathfrak{M} + R_s)x^j + (2\mathfrak{N} + 2R - sR_s)y^j\right]. \tag{7.88}$$

By using (7.80) we have

$$R_s = (m-1)R_{1s} + (r^2 - s^2)R_{2s} - 2sR_2. \tag{7.89}$$

From which together with (7.85) we have

$$2\mathfrak{M} + R_s = (m+1)R_{1s} + 3(r^2 - s^2)R_{2s} + 2(m+1)R_4 := \vartheta. \tag{7.90}$$

By (7.80), (7.86), (7.89) and (7.90),

$$2\mathfrak{N} + 2R - sR_s = -s\vartheta. \tag{7.91}$$

Substituting (7.90) and (7.91) into (7.88), we obtain the following formula for Ξ:

$$\Xi_j = -\frac{\vartheta}{3}(ux^j - sy^j) \tag{7.92}$$

where ϑ is given in (7.90). Hence we have the following:

Lemma 7.2.4 *$F(x, y) = |y|\phi\left(|x|, \frac{\langle x, y\rangle}{|y|}\right)$ has vanishing Ξ-curvature if and only if $\vartheta = 0$ where ϑ is given in (7.90).*

7.2.3 Constant (or Isotropic) Flag Curvature Equation

Now we are going to find partial differential equations that characterize spherically symmetric Finsler metrics of isotropic flag curvature. We need the following (see Proposition 8.1.1 or Theorem 6.1):

Lemma 7.2.5 *Let $F(x, y) = |y|\phi\left(|x|, \frac{\langle x, y\rangle}{|y|}\right)$ be a spherically symmetric Finsler metric on $\mathbb{B}^m(r_\mu)$. Then F is of scalar curvature if and only if ϕ satisfies*

$$\begin{aligned} R_2 := &\, 2Q(2Q - sQ_s) + \frac{1}{r}(2Q_r - sQ_{rs} - rQ_{ss}) \\ &+ (r^2 - s^2)(2QQ_{ss} - Q_s^2) = 0, \end{aligned} \tag{7.93}$$

where Q is given in (3.2).

Let $F = u\phi(r, s)$ be a spherically symmetric Finsler metric. We are going to find a necessary condition on ϕ for F to be of scalar (flag) curvature with flag curvature $K(x, y)$. A simple calculation gives the following formula:

$$F^2\delta^i_j - FF_{y^j}y^i = u^2\phi^2\delta^{ij} - u\phi\phi_s x^j y^i - \phi(\phi - s\phi_s)y^i y^j. \tag{7.94}$$

By (7.78) and (7.94), we see that (7.72) is equivalent to

$$\begin{aligned} u^2(R_1 - K\phi^2)\delta^{ij} + u^2R_2x^ix^j + uR_3x^iy^j + u(R_4 + K\phi\phi_s)x^jy^i \\ + [R_5 + K\phi(\phi - s\phi_s)]\, y^iy^j = 0. \end{aligned} \tag{7.95}$$

Therefore we have the following ([28], proof of Proposition 3.2):

Lemma 7.2.6 *Let $F(x, y) = |y|\phi\left(|x|, \frac{\langle x, y\rangle}{|y|}\right)$ be a spherically symmetric Finsler metric on $\mathbb{B}^m(r_\mu)$. Suppose that F is of scalar (flag) curvature with flag curvature $K(x, y)$. Then*

$$R_1 = K\phi^2, \qquad R_4 = -K\phi\phi_s. \tag{7.96}$$

Assume that $F = |y|\phi\left(|x|, \frac{\langle x, y\rangle}{|y|}\right)$ is of scalar curvature. Using (7.93) and (7.90), we obtain

$$\vartheta = (m+1)(R_{1s} + 2R_4). \tag{7.97}$$

By a straightforward computation one has

$$R_{1s} = 2PP_s - \frac{1}{r}(P_r + sP_{rs} + rP_{ss}) + 2Q_s\left[1 + sP + (r^2 - s^2)P_s\right]$$
$$+ 2Q\left[P - sP_s + (r^2 - s^2)P_{ss}\right]$$

where we have made use of (7.73). Plugging this and (7.76) into (7.97) yields

$$\vartheta = 3(m+1)\left[\frac{1}{r}P_r - \frac{s}{r}P_{rs} - P_{ss} + 2Q(P - sP_s) + 2(r^2 - s^2)QP_{ss}\right].$$

Together with Lemmas 7.2.5, 7.2.6, 7.2.2 and 7.2.4 we obtain [57]

Theorem 7.2.1 *Let $F(x, y) = |y|\phi\left(|x|, \frac{\langle x, y\rangle}{|y|}\right)$ be a spherically symmetric Finsler metric on $\mathbb{B}^m(r_\mu)$. Then F is of isotropic flag curvature (or constant flag curvature when $m \geq 3$) if and only if*

$$2Q(2Q - sQ_s) + \frac{1}{r}(2Q_r - sQ_{rs} - rQ_{ss}) + (r^2 - s^2)(2QQ_{ss} - Q_s^2) = 0, \tag{7.98}$$

$$\frac{1}{r}P_r - \frac{s}{r}P_{rs} - P_{ss} + 2Q(P - sP_s) + 2(r^2 - s^2)QP_{ss} = 0, \tag{7.99}$$

where P and Q are given in (3.3) *and* (3.2). *In this case, the flag curvature of F satisfies* (7.96).

Since $m(\geq 3)$-dimensional Finsler metrics with isotropic flag curvature are of constant (flag) curvature, based on Theorem 7.2.1, we obtain the following:

Theorem 7.2.2 *Let $F(x, y) = |y|\phi\left(|x|, \frac{\langle x, y\rangle}{|y|}\right)$ be a spherically symmetric Finsler metric on $\mathbb{B}^m(r_\mu)$ of dimension $m \geq 3$. Then F is of constant flag curvature if and only if* (7.98) *and* (7.99) *hold. In this case, the flag curvature K of F satisfies* (7.96).

Remark 7.2.1 It is easy to prove that a two dimensional spherically symmetric Finsler metric has constant flag curvature K if and only if (7.98), (7.99) and the first equation of (7.96) holds from (7.95). As a consequence of Theorems 7.2.1 and 7.2.2, by taking $Q = 0$ in (7.99), we obtain the following result, weakening Zhou's condition P [85].

Corollary 7.2.1 *Let $F(x, y) = |y|\phi\left(|x|, \frac{\langle x, y\rangle}{|y|}\right)$ be a projective spherically symmetric Finsler metric on $\mathbb{B}^m(r_\mu)$. Then F is of isotropic flag curvature if and only if*

$$\frac{1}{r}(P - sP_s)_r + \frac{1}{s}(P - sP_s)_s = 0 \tag{7.100}$$

where P is given in (3.3)*. In particular, if the dimension $m \geq 3$, then F is of constant flag curvature if and only if* (7.100) *holds. In this case, the flag curvature of F satisfies* (7.96).

7.2.4 Local Projectively Flat Metrics of Constant Flag Curvature

Let us consider Douglas spherically symmetric metric $F(x, y) = |y|\phi\left(|x|, \frac{\langle x, y\rangle}{|y|}\right)$. Then $Q = a(r) + b(r)s^2$ (see (4.11) or [54]). Assume that F is of scalar curvature. By using (7.98) or (7.93) we have [38]

$$b(r) = \frac{2ra(r)^2 + a'(r)}{r - 2a(r)r^3}.$$

Take a look at the special case: when $a(r) = -\frac{1}{r}$,

$$Q(r, s) = -\frac{\Psi}{r^3} \tag{7.101}$$

where Ψ is defined in (7.171) [56]. We put

$$T := P - sP_s. \tag{7.102}$$

Then (7.99) is equivalent to

$$\frac{1}{2r}T_r + \frac{1}{s}\left(\frac{1}{2} - \Psi Q\right)T_s = -QT. \tag{7.103}$$

The characteristic equation of the quasi-linear PDE (7.103) is

$$\frac{dr}{\frac{1}{2r}} = \frac{ds}{\frac{1}{s}(\frac{1}{2} - \Psi Q)} = \frac{dT}{-QT}. \tag{7.104}$$

It follows that

$$\frac{r\Psi}{r+4\Psi}=c_1,\qquad \frac{T}{\sqrt{\frac{r}{r+4\Psi}}}=c_2$$

are independent integrals of (7.104). Hence the solution of (7.103) is

$$T=f\left(\frac{r\Psi}{r+4\Psi}\right)\sqrt{\frac{r}{r+4\Psi}},\tag{7.105}$$

where f is any continuously differentiable function. Let us consider the special solution of (7.103) in the form $T=c\sqrt{\frac{r}{r+4\Psi}}$ where c is constant. In this case

$$P=g(r)s+\frac{c\sqrt{r(r+4r^2-4s^2)}}{r(1+4r)},\tag{7.106}$$

where we have used (7.102). Let

$$\rho:=\sqrt{r(r+4r^2-4s^2)},\qquad \mu:=1+4r.\tag{7.107}$$

Then (7.106) simplifies to

$$P=g(r)s+\frac{c\rho}{r\mu}.\tag{7.108}$$

Now we determine $g(r)$ and c in (7.108) using our necessary condition (7.168). By direct calculations one obtains

$$\rho_s=-\frac{4rs}{\rho},\qquad \rho_r=\frac{r+6r^2-2s^2}{\rho}.\tag{7.109}$$

Thus

$$P_r=g's+\frac{2c}{\rho r\mu^2}(12rs^2-4r^3-\Psi),\tag{7.110}$$

$$P-sP_s=\frac{cr}{\rho},\qquad 2P-sP_s=gs+\frac{2c}{\rho\mu}(r\mu-2s^2),\tag{7.111}$$

$$sP_{ss}=-\frac{4cr^2}{\rho^3}s,\qquad P+sP_s=2gs+c\frac{r\mu-8s^2}{\rho\mu},\tag{7.112}$$

$$2P_r-sP_{rs}=g's+\frac{4c}{\mu^2\rho^3}(24r^3s^2-24rs^4-16r^5+6r^2s^2-8r^4-2s^4-r^3).\tag{7.113}$$

Using (7.101), we obtain

$$Q_s = \frac{2s}{r^3}, \quad Q_{ss} = \frac{2}{r^3}, \quad Q_s - sQ_{ss} = 0, \tag{7.114}$$

$$2Q - sQ_s = -\frac{2}{r}, \quad 2Q_r - sQ_{rs} = \frac{2}{r^2}. \tag{7.115}$$

Using (7.110), (7.111), (7.112), (7.113), (7.114) and (7.115), we compute the terms in (7.168) as follows.

$$\begin{aligned} A = & \frac{8crs}{\mu^2\rho^3}[4(4r-1)s^4 - r\mu(8r-1)s^2 + r^3\mu^2] + \frac{2c^2r}{\mu} \\ & + \frac{2cs}{\rho\mu}[r\mu(rg-2) + 8s^2] + (g' - \frac{2}{r^2} - 4g)s^2, \end{aligned} \tag{7.116}$$

$$\begin{aligned} B = & \frac{-c}{\mu^2\rho}[8(3+4r)s^4 - 2r\mu(8r+5)s^2 + r^2\mu^2\kappa] \\ & + (4g - g' + \frac{2}{r^2})s^3 - 2(1 + r\kappa g)s, \end{aligned} \tag{7.117}$$

$$\Phi = \frac{\rho\mu(rg-2)s + 2cr(r\mu - 2s^2)}{r\kappa\rho\mu + (rg-2)\rho\mu s^2 + cs\rho^2}, \tag{7.118}$$

where

$$\kappa := 1 + 2r. \tag{7.119}$$

By (7.116), (7.117) and (7.118), (7.168) holds if and only if

$$\frac{4cr}{\mu^2}X_4 s^4 + \frac{4\rho}{\mu}X_3 s^3 + cr^2 X_2 s^2 - r\rho X_3 s - cr^3 X_1 = 0, \tag{7.120}$$

where

$$X_1 := 3r\kappa\mu g + 2(10r+3) - 2c^2r, \tag{7.121}$$

$$X_2 := 2r^2\mu g^2 + 2r(24r+13)g - r\mu g' + \frac{2}{r\mu}(80r^2 + 30r + 1) - \frac{16c^2r}{\mu}, \tag{7.122}$$

$$X_3 := 2r(\mu - 2c^2r)g + 2r^2\kappa\mu g^2 - r\kappa\mu g' + \frac{2\mu}{r} - \frac{8c^2r}{\mu}, \tag{7.123}$$

$$X_4 := r\mu[\mu g' - 2r\mu g^2 - 2(12r+7)g] - \frac{2}{r}(40r^2 + 18r + 1) + 8c^2r. \tag{7.124}$$

Note that $r > 0$. By using (7.107) and (7.120), we have

$$X_j = X_j(r) = 0, \quad j = 1, \cdots, 4. \tag{7.125}$$

Take $j = 1$. Then we get

$$g = \frac{2(c^2 r - 10r - 3)}{3r\mu\kappa}. \tag{7.126}$$

In (7.125), take $j = 3$ and 4, then

$$\begin{aligned} 0 &= \frac{\kappa}{\mu} X_4 + X_3 \\ &= [\mu - 2c^2 r - \kappa(12r + 7)] \frac{4r(c^2 r - 10r - 3)}{3r\mu K} - \frac{8}{\mu}(20r^2 + 15r + 3 - 2c^2 r^2). \end{aligned}$$

Furthermore, it simplifies to $2r^2(c^2 - 1)(c^2 - 4) = 0$. It follows that

$$c = \pm 1, \qquad c = \pm 2. \tag{7.127}$$

Plugging (7.126) into (7.108) yields

$$P = \frac{2(c^2 r - 10r - 3)}{3r\mu\kappa} s + \frac{c\rho}{r\mu}. \tag{7.128}$$

Combining this with the first equation of (7.109) yields

$$P_s = g - \frac{4cs}{\mu\rho}. \tag{7.129}$$

By (7.73), (7.110), (7.126), (7.128) and (7.129), we obtain

$$R_1 = \frac{c^2 - 4}{3} \left\{ \frac{1}{\mu} + \frac{4c\rho}{r\kappa\mu^2} s - \frac{4[12r^2 - (c^2 - 10)r + 3]}{3r\mu^2\kappa^2} s^2 \right\}.$$

Take $c = \pm 1$. Then

$$R_1 = -\left[\frac{1}{\mu} \pm \frac{4\rho}{r\kappa\mu^2} s - \frac{4(4r^2 + 3r + 1)}{r\kappa^2\mu^2} s^2 \right]. \tag{7.130}$$

By (7.73) and (7.130), we see that the first equation of (7.96) is equivalent to

$$P - \frac{sP_r + rP_s}{r} + 2Q(1 + sP + \Psi P_s) = -\phi^2,$$

where

$$\phi := \sqrt{\frac{1}{\mu} \pm \frac{4\rho}{r\kappa\mu^2}s - \frac{4(4r^2+3r+1)}{r\kappa^2\mu^2}s^2}. \tag{7.131}$$

By straightforward calculations, we have

$$\frac{1}{2r}\frac{r\phi_{ss} - \phi_r + s\phi_{rs}}{\phi - s\phi_s + \Psi\phi_{ss}} = -\frac{\Psi}{r^3}$$

and

$$\frac{r\phi_s + s\phi_r}{2r\phi} + \frac{\Psi}{r^3\phi}(s\phi + \Psi\phi_s) = -\frac{2(3r+1)}{r\mu\kappa}s \pm \frac{\rho}{r\mu}$$

that is, $\phi = \phi(r, s)$ satisfies the first equation of (7.96) (with $K = -1$), (7.98), (7.99), (3.2) and (3.3). By Theorem 7.2.8, $F = |y|\phi(|x|, \frac{\langle x,y\rangle}{|y|})$ is of constant curvature $K = -1$. Furthermore, (7.101) tells us F is locally projectively flat.

Thus we prove the following [57]:

Theorem 7.2.3 *The following spherically symmetric Finsler metrics are of constant curvature $K = -1$*

$$F(x, y) := \sqrt{\frac{|y|^2}{1+4|x|} \pm \frac{4\xi\langle x, y\rangle}{|x|(1+2|x|)(1+4|x|)^2} - \frac{4(4|x|^2+3|x|+1)\langle x, y\rangle^2}{|x|(1+2|x|)^2(1+4|x|)^2}}, \tag{7.132}$$

where

$$\xi := \sqrt{|x|[|x||y|^2 + 4(|x|^2|y|^2 - \langle x, y\rangle^2)]}. \tag{7.133}$$

Moreover, F is locally projectively flat.

In (7.127), taking $c = \pm 2$, we have the following:

Theorem 7.2.4 *The following spherically symmetric Finsler metrics are of constant curvature $K = 0$*

$$F(x, y) := \frac{[\xi(1+2|x|) \pm 2|x|\langle x, y\rangle]^2}{\xi|x|(1+4|x|)^2}, \tag{7.134}$$

where ξ is given in (7.133). *Moreover, F is locally projectively flat.*

In Theorem 7.2 of [34], author claims that on a convex domain $\mathscr{U} \subset \mathbb{R}^n$, a spherically symmetric Finsler metric F is locally projectively flat with vanishing flag curvature if and only if F is given in (1.2). Actually, we have proved (7.134) is also a locally projectively flat spherically symmetric Finsler metric with constant flag curvature $K = 0$ which differs from Finsler metric (1.2).

Now we take a look at another special case: when $a(r) = -2$,

$$Q(r, s) = -2 + \frac{8s^2}{1 + 4r^2}. \tag{7.135}$$

It follows that

$$\frac{(1 + 4r^2)\Psi}{1 + 4\Psi} = c_1, \qquad \frac{T}{\sqrt{\frac{1+4r^2}{1+4\Psi}}} = c_2$$

are independent integrals of (7.104). Hence the solution of (7.103) is

$$T = f\left(\frac{(1 + 4r^2)\Psi}{1 + 4\Psi}\right)\sqrt{\frac{1 + 4r^2}{1 + 4\Psi}},$$

where f is any continuously differentiable function. Let us consider the special solution of (7.103) in the form $T = c\sqrt{\frac{1+4r^2}{1+4\Psi}}$ where c is constant. In this case

$$P = g(r)s + c\rho, \quad P_s = g(r) + c\rho_s \tag{7.136}$$

where we have used (7.102) and

$$\rho := \sqrt{\frac{1 + 4\Psi}{1 + 4r^2}}. \tag{7.137}$$

Now we determine $g(r)$ and c in (7.136) using our necessary condition (7.168). By direct calculations one obtains

$$\rho_s = -\frac{4s}{\rho\mu}, \quad \rho_r = \frac{16rs^2}{\rho\mu^2}, \quad \mu := 1 + 4r^2. \tag{7.138}$$

Thus

$$P_r = g's + \frac{16crs^2}{\rho\mu^2}, \quad P - sP_s = \frac{c}{\rho}, \tag{7.139}$$

$$2P - sP_s = gs + \frac{2c(\mu - 2s^2)}{\rho\mu}, \quad sP_{ss} = -\frac{4cs}{\rho^3\mu}, \tag{7.140}$$

$$P + sP_s = 2gs + \frac{c(\mu - 8s^2)}{\rho\mu}, \quad 2P_r - sP_{rs} = g's - \frac{64cr}{(\rho\mu)^3}s^4. \tag{7.141}$$

Using (7.135), we obtain

$$Q_s = \frac{16s}{\mu}, \quad Q_{ss} = \frac{16}{\mu}, \quad Q_s - sQ_{ss} = 0, \tag{7.142}$$

$$2Q - sQ_s = -4, \quad 2Q_r - sQ_{rs} = 0. \tag{7.143}$$

Using (7.140), (7.141), (7.142), (7.143) and (7.139), we compute the terms in (7.168) as follows.

$$A = g's^2 - \frac{32r}{\mu}s^2 - \frac{32r^2s^2}{\mu}g + 2g\frac{crs}{\rho} + 2rc^2 + (I) \tag{7.144}$$

where

$$(I) := \frac{4crs}{\mu\rho^3} + \frac{16crs}{\mu\rho^3}\Psi - \frac{64cr^3s^3}{\mu^2\rho^3} - \frac{32cr^3s}{\mu}\rho = \frac{2crs}{\mu^2\rho}\left[2\mu(1-4r^2) + 32r^2s^2\right]. \tag{7.145}$$

Plugging (7.145) into (7.144) yields

$$A = \frac{2crs}{\mu^2\rho}\left[2\mu(1-4r^2) + \mu^2 g + 32r^2s^2\right] + 2rc^2 + \left[g' - \frac{32r}{\mu}(1+r^2g)\right]s^2. \tag{7.146}$$

By using (7.142) and (7.143) we have

$$B = 4rsQ(1+sP) - s^2P_r - r(1-2Q\Psi)(P+sP_s) = (II) + (III) \tag{7.147}$$

where

$$\begin{aligned} (II) : &= 4rsQ(1+gs^2) - g's^3 - 2rg(1-2Q\Psi)s \\ &= 4rsQ - 2r\mu gs + \frac{32r^3}{\mu}gs^3 - g's^3, \end{aligned} \tag{7.148}$$

$$\begin{aligned} (III) : &= 4crs^2\rho Q - \frac{16cr}{\mu^2\rho}s^4 - cr(1-2Q\Psi)\frac{\mu - 8s^2}{\mu\rho} \\ &= -\frac{cr\rho}{\mu}(\mu^2 - 16r^2s^2). \end{aligned} \tag{7.149}$$

Substituting (7.148) and (7.149) into (7.147), we have

$$B = -\frac{cr\rho}{\mu}(\mu^2 - 16r^2s^2) - 2r(4+\mu g)s + \left[\frac{32r}{\mu}(1+r^2g) - g'\right]s^3. \tag{7.150}$$

By using (7.136), (7.140) and (7.143) we obtain

$$\begin{aligned}\Phi =& \frac{2P - sP_s - 4s}{1 + 4\Psi + sP} \\ =& \frac{gs + \frac{2c(\mu - 2s^2)}{\mu\rho} - 4s}{\mu - 4s^2 + s(gs + c\rho)} = \frac{\mu\rho(g-4)s + 2c(\mu - 2s^2)}{\mu^2\rho + \mu\rho(g-4)s^2 + c\mu\rho^2 s}.\end{aligned} \tag{7.151}$$

By (7.146), (7.150) and (7.151), (7.168) holds if and only if

$$\frac{c}{\mu\rho}X_2 s^2 + X_1 s + \frac{cr}{\rho}X_0 = 0 \tag{7.152}$$

where

$$X_0 := 2c^2 - 8 - 3\mu g, \tag{7.153}$$

$$X_1 := \mu g' + 4c^2 rg - 2r\mu g^2, \tag{7.154}$$

$$X_2 := 8r(4 - c^2) + 8r(1 + 6r^2)g + 2r\mu g^2 - \mu g'. \tag{7.155}$$

From which together with (7.137) and (7.152) we obtain the following

$$X_j = X_j(r) = 0, \quad j = 0,\ 1,\ 2. \tag{7.156}$$

It follows that

$$0 = X_1 + X_2 = 4r\left[c^2 + 2(1 + 6r^2)\right]g + 8r(4 - c^2)$$

where we have used (7.154) and (7.155). Note that $r > 0$. Hence we have

$$g = \frac{2(c^2 - 4)}{c^2 + 2(1 + 6r^2)}. \tag{7.157}$$

By (7.153) and (7.156), we have $g = \frac{2(c^2-4)}{3\mu}$. Combine this with (7.157) we obtain

$$c = \pm 2, \quad c = \pm 1. \tag{7.158}$$

Take $c = \pm 2$. Then $g = 0$. In this case, we have

$$\begin{aligned}&P^2 - \frac{1}{r}(sP_r + rP_s) + 2Q[1 + sP + (r^2 - s^2)P_s] \\ &= c^2 - 4 - \mu g + 2c\rho g s + \left[g^2 + \frac{16r^2}{\mu}g - \frac{1}{r}g' - \frac{4}{\mu}(c^2 - 4)\right] = 0\end{aligned}$$

where we have used (7.135), (7.136) and the first equation of (7.139). It is impossible to solve ϕ by using the Eq. (7.72) and the first equation of (7.96). Now we are going to solve ϕ by using another approach.

From (7.136), (7.140) and (7.143), we obtain

$$P = 2\rho, \quad 2P - sP_s = \pm\frac{4(\mu - 2s^2)}{\mu\rho}, \quad 2Q - sQ_s = -4. \tag{7.159}$$

Substituting (7.159) into (7.196) yields

$$\begin{aligned}(\log\phi)_s =& \frac{\pm 4(\mu - 2s^2) - 4s\mu\rho}{\mu\rho\left[1 + 4(r^2 - s^2) \pm 2s\rho\right]} \\ =& \frac{(-4s \pm 2\rho)\rho \pm 2}{\rho\left[\mu - 4s^2 \pm 2s\rho\right]} = \frac{-4s \pm 2\rho}{\rho(\mu\rho \pm 2s)} \pm \frac{2}{\rho^2(\mu\rho \pm 2s)}\end{aligned} \tag{7.160}$$

where we have used the fact $\rho^2\mu = \mu - 4s^2$. It follows that

$$\begin{aligned}\phi =& e^{\log\phi} \\ =& t_1(r) e^{2\int \frac{(-2s\pm\rho)\rho\pm 1}{\rho^2(\mu\rho\pm 2s)} ds} \\ =& t_2(r)\frac{(\mu\rho \pm 2s)^2}{\rho} = t(r)\frac{(\mu\rho \pm 2s)^2}{\mu^2\rho}.\end{aligned} \tag{7.161}$$

In particular,

$$\log\phi(r, 0) = \log\left[t(r)\rho(r, 0)\right] = \log t(r) \tag{7.162}$$

where we have made use of (7.137). Using (7.135), (7.136), (7.142) and (7.182), we have

$$Q = -2\rho^2, \quad P_s = \mp\frac{8s}{\mu\rho}, \quad Q_s = \frac{16s}{\mu}, \quad E = \rho(\rho\mu \pm 2s). \tag{7.163}$$

Together with the first equation of (7.159) we have the following:

$$\begin{aligned}& P_s + 2P^2 + 2sPQ + sQ_s + 2\Psi(PQ_s - QP_s) \\ =& \mp\frac{8s}{\mu\rho} + 8\rho^2 \mp 8s\rho^3 + \frac{16}{\mu}s^2 \pm 2\Psi\frac{16\rho}{\mu}s.\end{aligned} \tag{7.164}$$

Plugging (7.164) and the last equation of (7.163) into (7.197) yields

$$(\log\phi)_r = \frac{2r}{\rho(\mu\rho \pm \pm 2s)}\left(\mp\frac{4s}{\mu\rho} + 4\rho^2 \mp 4s\rho^3 + \frac{8}{\mu}s^2 \pm \Psi\frac{16\rho}{\mu}s\right).$$

Together with (7.162) we obtain

$$(\log \mu)' = \frac{8r}{\mu} = (\log \phi)_r|_{s=0} = \left[\log \phi(r, 0)\right]' = [t(r)]'.$$

It follows that $\log \frac{t(r)}{\mu} = \lambda = constant$. We may assume that $\lambda = 0$. Hence we have

$$t(r) = \mu = 1 + 4r^2.$$

Plugging this into (7.161) yields $\phi(r, s) = \frac{(\mu\rho \pm 2s)^2}{\mu\rho}$. Define ξ by

$$\xi(x, y) := |y|\rho\left(|x|, \frac{\langle x, y\rangle}{|y|}\right) = \sqrt{\frac{(1 + 4|x|^2)|y|^2 - 4\langle x, y\rangle^2}{1 + 4|x|^2}}. \tag{7.165}$$

Then

$$F(x, y) = |y|\phi\left(|x|, \frac{\langle x, y\rangle}{|y|}\right) = \frac{\left[\xi(1 + 4|x|^2) \pm 2\langle x, y\rangle\right]^2}{\xi(1 + 4|x|^2)}. \tag{7.166}$$

By straightforward calculations, we have

$$\frac{1}{2r}\frac{r\phi_{ss} - \phi_r + s\phi_{rs}}{\phi - s\phi_s + \Psi\phi_{ss}} = -2\frac{1 + 4\Psi}{1 + 4r^2},$$

$$\frac{r\phi_s + s\phi_r}{2r\phi} + \frac{\Psi}{r^3\phi}(s\phi + \Psi\phi_s) = \pm 2\sqrt{\frac{1 + 4\Psi}{1 + 4r^2}}$$

and

$$\phi - s\phi_s > 0, \quad \phi - s\phi_s + \Psi\phi_{ss} > 0,$$

that is, $\phi = \phi(r, s)$ satisfies the first equation of (7.96) (with $K = 0$), (7.98), (7.99), (3.2) and (3.3). By Theorem 7.2.1, (7.166) is of constant curvature $K = 0$. Furthermore, (7.135) tells us F is locally projectively flat. Thus we prove the following:

Theorem 7.2.5 *The following spherically symmetric Finsler metrics are locally projectively flat*

$$F(x, y) = \frac{\left[\xi(1 + 4|x|^2) \pm 2\langle x, y\rangle\right]^2}{\xi(1 + 4|x|^2)}, \tag{7.167}$$

where ξ is given in (7.165). *Moreover, F is of constant flag curvature $K = 0$.*

In (7.158), taking $c = \pm 1$, we have the following:

Theorem 7.2.6 *The following spherically symmetric Finsler metrics are locally projectively flat*

$$F(x, y) = \sqrt{|y|^2 \pm \frac{4\langle x, y\rangle \xi}{1 + 4|x|^2} - \frac{16|x|^2\langle x, y\rangle^2}{(1 + 4|x|^2)^2}},$$

where ξ is given in (7.165). *Moreover, F is of constant flag curvature $K = -1$.*

In Theorem 7.3 of [34], author claims that on a convex domain $\mathscr{U} \subset \mathbb{R}^n$, a spherically symmetric Finsler metric F is locally projectively flat with constant flag curvature $K = -1$ if and only if F is given in (7.3). Actually, we have proved the Finsler metric in Theorem 7.2.6 is also a locally projectively flat spherically symmetric Finsler metric with constant flag curvature $K = -1$ which differs from Finsler metric (7.3).

A natural task for us is to classify all locally projectively flat spherically symmetrics Finsler metric with constant flag curvature.

7.2.5 Integrable Condition

We establish the Proposition and Lemma required in the proof of Theorems 7.2.3 and 7.2.5 respectively.

Proposition 7.2.1 *Let $P(r, s)$ and $Q(r, s)$ be differentiable functions. If there exists a differentiable function $\phi = \phi(r, s)$ such that* (3.2) *and* (3.3) *hold, then*

$$A + B\Phi = 0, \tag{7.168}$$

where

$$A := s[2P_r - sP_{rs} + s(2Q_r - sQ_{rs})] + 2rsQP_{ss}\Psi - 2rQ_s(s + r^2P)$$
$$+ rs(Q_s - sQ_{ss})(1 - 2Q\Psi) + 2rP(P - sP_s) - rsP_{ss}, \tag{7.169}$$

$$B := 4rsQ(1 + sP) - s[sP_r - (2Q_r - sQ_{rs})\Psi]$$
$$+ r(1 - 2Q\Psi)[(Q_s - sQ_{ss})\Psi - (P + sP_s)], \tag{7.170}$$

$$\Phi := \frac{2P - sP_s + s(2Q - sQ_s)}{1 - \Psi(2Q - sQ_s) + sP}, \qquad \Psi := r^2 - s^2. \tag{7.171}$$

In the first we show the following

Lemma 7.2.7 *Let $P(r, s)$ and $Q(r, s)$ be differentiable functions. If there exists a differentiable function $\phi = \phi(r, s)$ such that* (3.2) *and* (3.3) *hold, then*

$$s^2\Phi_r + rs\Phi_s - r\Phi = sW_s - W, \tag{7.172}$$

where

$$W := \frac{s\phi_r + r\phi_s}{\phi}, \quad \Phi := (\log\phi)_s. \tag{7.173}$$

Moreover, Φ satisfies (7.171).

Proof Define

$$U := \frac{s\phi + \Psi\phi_s}{\phi} = s + \Psi(\log\phi)_s = s + \Psi\Phi. \tag{7.174}$$

It follows that

$$\Phi = \frac{U - s}{\Psi}. \tag{7.175}$$

By (7.173), we have $(\log\phi)_r = \frac{1}{s}(W - r\Phi)$. Therefore $(\log\phi)_{rs} = \frac{1}{s}(W_s - r\Phi_s) - \frac{1}{s^2}(W - r\Phi)$. Together with (7.173) we obtain

$$\Phi_r = (\log\phi)_{sr} = (\log\phi)_{rs} = \frac{1}{s}(W_s - r\Phi_s) - \frac{1}{s^2}(W - r\Phi). \tag{7.176}$$

Equation (7.172) follows from (7.176). Now we compute Φ. By (3.2), (3.3), (7.173) and (7.174) we have

$$P = -UQ + \frac{W}{2r} \tag{7.177}$$

and

$$\begin{aligned} Q &= \frac{1}{2r}\frac{(s\phi_r + r\phi_s)_s - 2\phi_r}{(s\phi + \Psi\phi_s)_s} \\ &= \frac{1}{2r}\frac{(\phi W)_s - 2\phi_r}{(\phi U)_s} \\ &= \frac{1}{2r}\frac{W(\log\phi)_s + W_s - 2(\log\phi)_r}{U(\log\phi)_s + U_s} = \frac{1}{2rs}\frac{sW_s + (sW + 2r)\Phi - 2W}{U_s + U\Phi}. \end{aligned} \tag{7.178}$$

From (7.177), one has

$$W = 2r(P + QU). \tag{7.179}$$

Therefore

$$W_s = 2r(P + QU)_s. \tag{7.180}$$

Plugging (7.179) and (7.180) into (7.178) yields

$$Q = \frac{s(P_s + Q_s U + U_s Q) + [s(P + QU) + 1]\Phi - 2(P + QU)}{s(U_s + U\Phi)}. \tag{7.181}$$

It follows that

$$\begin{aligned} 0 &= s(P_s + Q_s U) + (sP + 1)\Phi - 2(P + QU) \\ &= s(P_s + Q_s U) + (sP + 1)\frac{U - s}{\Psi} - 2(P + QU), \end{aligned}$$

where we have used (7.175). This gives

$$U = \frac{s + r^2 P + \Psi(P - sP_s)}{1 - \Psi(2Q - sQ_s) + sP},$$

from which together with (7.175) we obtain (7.171).

Proof of Proposition 7.2.1 Define

$$E := 1 + sP - \Psi(2Q - sQ_s). \tag{7.182}$$

Then

$$E_r = sP_r - 2r(2Q - sQ_s) - \Psi(2Q_r - sQ_{rs}) \tag{7.183}$$

and

$$E_s = P + sP_s + 2s(2Q - sQ_s) - \Psi(Q_s - sQ_{ss}). \tag{7.184}$$

Using (7.182) and (7.171), we have

$$E\Phi = 2P - sP_s + s(2Q - sQ_s). \tag{7.185}$$

It follows that $(E\Phi)_r = 2P_r - sP_{rs} + s(2Q_r - sQ_{rs})$ and $(E\Phi)_s = P_s - sP_{ss} + 2Q - s^2 Q_{ss}$. Thus we obtain

$$E\Phi_r = 2P_r - sP_{rs} + s(2Q_r - sQ_{rs}) - [sP_r - 2r(2Q - sQ_s) - \Psi(2Q_r - sQ_{rs})]\Phi \tag{7.186}$$

and

$$E\Phi_s = P_s - sP_{ss} + 2Q - s^2 Q_{ss} - [P + sP_s + 2s(2Q - sQ_s) - \Psi(Q_s - sQ_{ss})]\Phi, \tag{7.187}$$

where we have made use of (7.183) and (7.184). Plugging (7.174) into (7.179) yields

$$\frac{W}{2r} = P + sQ + \Psi\Phi Q. \tag{7.188}$$

It follows that

$$\begin{aligned}\frac{W_s}{2r} &= P_s + sQ_s + Q - 2s\Phi Q + \Psi\Phi Q_s \\ &+ \frac{\Psi Q}{E}\{P_s - sP_{ss} + 2Q - s^2 Q_{ss} - [P + sP_s + 2s(2Q - sQ_s) - \Psi(Q_s - sQ_{ss})]\Phi\},\end{aligned}$$

where we have used (7.187). Together with (7.182) we have

$$\begin{aligned}\frac{EW_s}{2r} &= (P_s + sQ_s)E + Q(1 + sP) + Q\Psi(P_s - sP_{ss} + sQ_s - s^2 Q_{ss}) \\ &\quad + \Phi[\Psi Q_s E - 2sQ(1 + sP) - (P + sP_s)Q\Psi + (Q_s - sQ_{ss})Q\Psi^2].\end{aligned} \tag{7.189}$$

By (7.186) and (7.187), we obtain

$$\begin{aligned}& s^2 E\Phi_r + rsE\Phi_s - rE\Phi = s^2 \\ & \times \left\{2P_r - sP_{rs} + s(2Q_r - sQ_{rs}) - [sP_r - 2r(2Q - sQ_s) - \Psi(2Q_r - sQ_{rs})]\Phi\right\} \\ & + rs\left\{P_s - sP_{ss} + 2Q - s^2 Q_{ss} - [P + sP_s + 2s(2Q - sQ_s) - \Psi(Q_s - sQ_{ss})]\Phi\right\} \\ & - rE\Phi.\end{aligned} \tag{7.190}$$

It follows from (7.188) and (7.189) that

$$\begin{aligned}sEW_s - EW &= -2rE(P + sQ + Q\Psi\Phi) \\ &\quad + 2rs[(P_s + sQ_s)E + Q(1 + sP) + Q\Psi(P_s - sP_{ss} + sQ_s - s^2 Q_{ss})] \\ &\quad + 2rs\Phi[\Psi Q_s E - 2sQ(1 + sP) - (P + sP_s)Q\Psi + (Q_s - sQ_{ss})Q\Psi^2].\end{aligned} \tag{7.191}$$

By (7.190) and (7.191), (7.172) holds if and only if

$$T_1 + T_2\Phi = 0, \tag{7.192}$$

where

$$
\begin{aligned}
T_1 := &\ (I) + rs(P_s - sP_{ss} + 2Q - s^2Q_{ss}) + 2rE(P + sQ) \\
&- 2rs[(P_s + sQ_s)E + (1 + sP)Q + Q\Psi(P_s - sP_{ss} + sQ_s - s^2Q_{ss})] \\
= &\ (I) + rs(1 - 2Q\Psi)(P_s - sP_{ss} + sQ_s - s^2Q_{ss}) - 4rsQ(1 + sP) \\
&+ rs(1 + 2Q\Psi)(2Q - sQ_s) + 2r(P - sP_s + 2sQ - s^2Q_s)E,
\end{aligned}
\tag{7.193}
$$

where $(I) := s^2[2P_r - sP_{rs} + s(2Q_r - sQ_{rs})]$ and we have used (7.182). In (7.192),

$$
\begin{aligned}
T_2 := &\ -s^2[sP_r - 2r(2Q - sQ_s) - \Psi(2Q_r - sQ_{rs})] - rE \\
&- rs[P + sP_s + 2s(2Q - sQ_s) - \Psi(Q_s - sQ_{ss})] + 2rQ\Psi E \\
&- 2rs[\Psi Q_s E - 2sQ(1 + sP) - (P + sP_s)Q\Psi + (Q_s - sQ_{ss})Q\Psi^2] \\
= &\ 4rs^2Q(1 + sP) - s^2[sP_r - (2Q_r - sQ_{rs})\Psi] - r(1 + 2Q\Psi)E \\
&+ rs(1 - 2Q\Psi)[(Q_s - sQ_{ss})\Psi - (P + sP_s)] + 2r\Psi(2Q - sQ_s)E.
\end{aligned}
\tag{7.194}
$$

From (7.193) and (7.194), we obtain

$$
\begin{aligned}
T_1 + T_2\Phi = &\ (I) + sB\Phi + rs(1 - 2Q\Psi)(P_s - sP_{ss} + sQ_s - s^2Q_{ss}) \\
&- 4rsQ(1 + sP) + rs(1 + 2Q\Psi)(2Q - sQ_s) \\
&+ 2r(P - sP_s + 2sQ - s^2Q_s)[1 + sP - \Psi(2Q - sQ_s)] \\
&+ r[2\Psi(2Q - sQ_s) - (1 + 2Q\Psi)][2P - sP_s + s(2Q - sQ_s)] \\
= &\ s(A + B\Phi),
\end{aligned}
\tag{7.195}
$$

where A and B are defined in (7.169) and (7.170) respectively. Now (7.168) follows from (7.172), (7.192) and (7.195).

Lemma 7.2.8 *Let $P(r, s)$ and $Q(r, s)$ be differentiable functions. If there exists a differentiable function $\phi = \phi(r, s)$ such that* (3.2) *and* (3.3) *hold, then*

$$
(\log\phi)_s = \frac{2P - sP_s + s(2Q - sQ_s)}{E}, \tag{7.196}
$$

$$
(\log\phi)_r = \frac{r}{E}\Big[P_s + 2P^2 + 2sPQ + sQ_s + 2\Psi(PQ_s - QP_s)\Big] \tag{7.197}
$$

where E is defined in (7.178).

Proof The formula (7.196) is immediate from Lemma 7.2.7. Now we are going to show (7.197). By using (7.196), we have

$$U := s + \Psi(\log\phi)_s = \frac{s + r^2 P + \Psi(P - sP_s)}{E}. \tag{7.198}$$

Combining the first equation of (7.173) with (3.2), (3.3) and (7.198) we have

$$W = 2r(P + QU) = 2r\left[P + Q\frac{s + r^2 P + \Psi(P - sP_s)}{E}\right].$$

It follows that

$$\begin{aligned}(\log\phi)_r =& \frac{W - r(\log\phi)_s}{s}\\ =& \frac{2r}{s}\left[P + Q\frac{s + r^2 P + \Psi(P - sP_s)}{E}\right] - \frac{r}{s}\frac{2P - sP_s + s(2Q - sQ_s)}{E}\\ =& \frac{r}{sE}(I)\end{aligned} \tag{7.199}$$

where we have made use of (7.196) and (7.173), and

$$\begin{aligned}(I) :=& 2PE + 2Q\left[s + r^2 P + \Psi(P - sP_s)\right] - [2P - sP_s + s(2Q - sQ_s)]\\ =& s\left[P_s + 2P^2 + 2sPQ + sQ_s + \Psi(PQ_s - QP_s)\right].\end{aligned} \tag{7.200}$$

Plugging (7.200) into (7.199) yields (7.197).

Chapter 8
Spherically Symmetric W-Quadratic Metrics

Two Finsler metrics on a manifold are said to be *(pointwise) projectively related* (*projectively equivalent* in an alternative terminology in [13]), if they have the same geodesics as point sets.

The Weyl curvature is one of the fundamental quantities in Finsler geometry because it is a projective invariant. Namely, if two Finsler metrics F and $\widetilde{F}$ are projectively related, then F and $\widetilde{F}$ have same Weyl curvature.

A Finsler metric is said to be *W-quadratic* if it has quadratic Weyl curvature. In this chapter, we are going to study spherically symmetric W-quadratic Finsler metrics. In particular, we give a lot of new spherically symmetric Finsler metrics of quadratic Weyl curvature which are non-trivial in the sense that they are not of Weyl type or quadratic Riemannian curvature.

8.1 Finsler Metrics with Special Riemannian Curvature Properties

In Finsler geometry, there are several important quantities: the Riemannian curvature, the Ricci curvature and the Weyl curvature, etc. They are extensions of corresponding quantities in Riemannian geometry, hence they said to be *Riemannian*.

A Finsler metric is called a *Weyl metric* if it has vanishing Weyl curvature.

8.1.1 Weyl Curvature and Weyl Metrics

In this subsection, we are going to calculate a formula for the Weyl curvature of a spherically symmetric metric. In particular, we give the equation that characterizes

E. Guo, X. Mo, *The Geometry of Spherically Symmetric Finsler Manifolds*, SpringerBriefs in Mathematics, https://doi.org/10.1007/978-981-13-1598-5_8

spherically symmetric Weyl metrics. Let

$$A^i_j := R^i_j - \frac{Ric}{m-1}\delta^i_j. \tag{8.1}$$

where R^i_j and Ric denote the Riemannian curvature and Ricci curvature respectively. Then the *(projective) Weyl curvature* $W_y = W^i{}_j \frac{\partial}{\partial x^i} \otimes dx^j$ is defined by

$$W^i{}_j := A^i_j - \frac{1}{m+1}\sum_{k=1}^{m} \frac{\partial A^k_j}{\partial y^k} y^i. \tag{8.2}$$

A Finsler metric is called a *Weyl metric* if $W^i{}_j = 0$, that is, it has vanishing Weyl curvature. We have the following interesting result [42]: A Finsler metric has vanishing Weyl curvature if and only if it is of scalar curvature. It follows that the Weyl curvature gives a measure of the failure of a Finsler metric to be of scalar (flag) curvature.

Now we compute the Weyl curvature of a spherically symmetric metric $F(x, y) = |y|\phi\left(|x|, \frac{\langle x, y\rangle}{|y|}\right)$.

By using (7.81) and (7.84), we have

$$\frac{\partial}{\partial y^j} Ric = uR_s x^j + (2R - sR_s)y^j \tag{8.3}$$

$$\sum_i \frac{\partial R^i_j}{\partial y^i} = u\mathfrak{M}x^j + \mathfrak{N}y^j, \tag{8.4}$$

where $R_s = \frac{\partial R}{\partial s}$ and R, $\mathfrak{M}$, $\mathfrak{N}$ are defined in (7.80), (7.85), (7.86) respectively.

From (8.1), (8.3) and (8.4), we have

$$\begin{aligned}\sum_k \frac{\partial A^k_j}{\partial y^k} &= \sum_k \frac{\partial R^k_j}{\partial y^k} - \frac{1}{m-1}\frac{\partial Ric}{\partial y^j} \\ &= u\left(\mathfrak{M} - \frac{R_s}{m-1}\right)x^j + \left[\mathfrak{N} - \frac{1}{m-1}(2R - sR_s)\right]y^j.\end{aligned}$$

Plugging this and (8.1) into (8.2) yields

$$
\begin{aligned}
W^i{}_j =& R^i_j - \frac{Ric}{m-1}\delta^i_j - \frac{1}{m+1}\left\{u\left(\mathfrak{M} - \frac{R_s}{m-1}\right)x^j + \left[\mathfrak{N} - \frac{1}{m-1}(2R - sR_s)\right]y^j\right\}y^i \\
=& u^2R_1\delta^i_j + u^2R_2x^ix^j + uR_3x^iy^j + uR_4x^jy^i + R_5y^iy^j \\
&- \frac{u^2}{m-1}R\delta^i_j - \frac{u}{m+1}\left(\mathfrak{M} - \frac{R_s}{m-1}\right)x^jy^i - \frac{1}{m+1}\left[\mathfrak{N} - \frac{1}{m-1}(2R - sR_s)\right]y^iy^j \\
=& u^2W_1\delta^i_j + u^2W_2x^ix^j + uW_3x^iy^j + uW_4x^jy^i + W_5y^iy^j
\end{aligned}
\tag{8.5}
$$

where we have used (7.78) and

$$
W_1 := R_1 - \frac{R}{m-1} = -\frac{r^2 - s^2}{m-1}R_2, \tag{8.6}
$$

$$
W_2 := R_2, \tag{8.7}
$$

$$
W_3 := R_3 = -sR_2, \tag{8.8}
$$

$$
W_4 := R_4 - \frac{1}{m+1}\left(\mathfrak{M} - \frac{R_s}{m-1}\right) = -\frac{s}{m-1}R_2 - \frac{m-2}{m^2-1}(r^2 - s^2)R_{2s}, \tag{8.9}
$$

$$
\begin{aligned}
W_5 :=& R_5 - \frac{1}{m+1}\left[\mathfrak{N} - \frac{1}{m-1}(2R - sR_s)\right] \\
=& \frac{r^2}{m-1}R_2 + \frac{m-2}{m^2-1}s(r^2 - s^2)R_{2s}.
\end{aligned}
\tag{8.10}
$$

Form (8.5), (8.6), (8.7), (8.8), (8.9) and (8.10) we obtain the following:

Proposition 8.1.1 *Let $F(x, y) = |y|\phi\left(|x|, \frac{\langle x, y\rangle}{|y|}\right)$ be a spherically symmetric Finsler metric on $\mathbb{B}^m(r_\mu)$. Then F is a Weyl metric if and only if ϕ satisfies*

$$
R_2 := 2Q(2Q - sQ_s) + \frac{1}{r}(2Q_r - sQ_{rs} - rQ_{ss}) + (r^2 - s^2)(2QQ_{ss} - Q_s^2) = 0 \tag{8.11}
$$

where Q is given in (3.2).

8.1.2 W-Quadratic Metrics

A Finsler metric F with $W^i{}_j$ quadratic in y is said to be *W-quadratic* [13, 36]. Note that every Weyl metric must be W-quadratic metric.

First we give the following:

Proposition 8.1.2 *Let $Q(r,s)$ be a polynomial function with respect to s defined by*

$$
Q(r, s) = f_0(r) + f_1(r)s + \cdots + f_k(r)s^k \tag{8.12}
$$

where $f_k(r) \neq 0$. Then Q satisfies (8.11) *if and only if*

$$Q = f_0(r) + f_2(r)s^2 \tag{8.13}$$

and

$$2f_0^2 + \frac{1}{r}f_0' - f_2 + 2r^2 f_0 f_2 = 0. \tag{8.14}$$

Proof Case 1. $k \geq 1$. By using (8.12), we have

$$Q_r = \textstyle\sum_{j=0}^k f_j' s^j, \quad Q_s = \sum_{j=1}^k j f_j s^{j-1},$$
$$Q_{rs} = \textstyle\sum_{j=1}^k j f_j' s^{j-1}, \quad Q_{ss} = \sum_{j=2}^k j(j-1) f_j s^{j-2}.$$

It follows that

$$R_2 \equiv (4-k^2) f_k^2 s^{2k} \quad mod\ s^0, s^1, \cdots, s^{2k-1}.$$

First suppose that $R_2 = 0$. Then $k = 2$. Thus

$$Q(r,s) = f_0(r) + f_1(r)s + f_2(r)s^2. \tag{8.15}$$

It follows that

$$Q_r = f_0' + f_1' s + f_2' s^2, \quad Q_s = f_1 + 2f_2 s,$$
$$Q_{rs} = f_1' + 2f_2' s, \quad Q_{ss} = 2f_2.$$

Plugging these into (8.11) yields

$$R_2 = 4f_0^2 + \frac{2}{r}f_0' - r^2 f_1^2 - 2f_2 + 4r^2 f_0 f_2 + \left(6f_0 f_1 + \frac{1}{r}f_1'\right)s + 3f_1^2 s^2 + 2f_1 f_2 s^3. \tag{8.16}$$

We obtain

$$f_1(r) = 0 \tag{8.17}$$

and (8.14) holds. Substituting (8.17) into (8.15) yields (8.13).

Conversely, we suppose that (8.13) and (8.14) hold. Then $f_1(r) = 0$ and $k = 2$ from (8.12). Plugging these and (8.14) into (8.16) we obtain $R_2 = 0$.

Case 2. $k = 0$. This is an immediate conclusion of (8.15) and (8.16).

Proposition 8.1.3 *Let $F(x, y) = |y|\phi\left(|x|, \frac{\langle x, y\rangle}{|y|}\right)$ be a spherically symmetric Finsler metric on $\mathbb{B}^m(r_\mu)$ satisfying $Q(r,s) = f_0(r) + f_2(r)s^2$. Then the Weyl curvature of F is given by*

$$W^i{}_j = W^i_{jkl}(x)y^k y^l \tag{8.18}$$

where

$$W^i_{jkl}(x) = R_2(|x|)T^i_{jkl}(x) \tag{8.19}$$

where

$$R_2(r) = 4f_0^2(r) + \frac{2}{r}f_0'(r) - 2f_2(r) + 4r^2 f_0(r)f_2(r) \tag{8.20}$$

and

$$\begin{aligned} T^i_{jkl} =&\delta_{kl}x^i x^j + \frac{1}{m-1}\delta^i_j x^k x^l + \frac{|x|^2}{m-1}\delta^i_k \delta_{jl} \\ &- \delta_{jl}x^i x^k - \frac{|x|^2}{m-1}\delta^i_j \delta_{kl} - \frac{1}{m-1}\delta^i_l x^j x^k. \end{aligned} \tag{8.21}$$

In particular, F is W-quadratic.

Proof (8.15), (8.17) and (8.16) imply (8.20). Thus we have $R_{2s} = 0$. Substituting this into (8.6), (8.7), (8.8), (8.9) and (8.10), we have

$$(W_1, W_2, W_3, W_4, W_5) = R_2\left(-\frac{r^2 - s^2}{m-1}, 1, -s, -\frac{s}{m-1}, \frac{r^2}{m-1}\right).$$

Plugging this into (8.5) yields

$$\begin{aligned} W^i{}_j =&R_2\Big[u^2\left(-\frac{r^2 - s^2}{m-1}\right)\delta^i_j + u^2 x^i x^j + u(-s)x^i y^j \\ &+ u\left(-\frac{s}{m-1}\right)x^j y^i + \frac{r^2}{m-1}y^i y^j\Big] \\ =&R_2\Big[-\frac{1}{m-1}\left(|x|^2|y|^2 - \langle x, y\rangle^2\right)\delta^i_j + |y|^2 x^i x^j \\ &- \langle x, y\rangle x^i y^j - \frac{\langle x, y\rangle}{m-1}x^j y^i + \frac{|x|^2}{m-1}y^i y^j\Big] \\ =&R_2(r)T^i_{jkl}(x)y^k y^l = W^i_{jkl}(x)y^k y^l. \end{aligned}$$

Thus we have proved Proposition 8.1.3.

8.1.3 New Finsler Metrics with Vanishing (or Quadratic) Weyl Curvature

Now we study spherically symmetric Finsler metrics satisfying $Q(r,s) = f_0(r)$ where Q is given in (3.2). It follows from (3.2) that

$$r\phi_{ss} - \phi_r + s\phi_{rs} = 2rf_0(r)\Big[\phi - s\phi_s + (r^2 - s^2)\phi_{ss}\Big]. \tag{8.22}$$

Lemma 8.1.1 *For $s > 0$, the general solution ϕ of (8.22) is given by*

$$\phi(r,s) = s\cdot h(r) - sg(r)\int_{s_0}^{s}\sigma^{-2}f\Big((r^2-\sigma^2)g^2(r)\Big)\,d\sigma \tag{8.23}$$

where $s_0 \in (0, s]$ and

$$g(r) := e^{-\int 2rf_0(r)dr}. \tag{8.24}$$

Proof Note that $s > 0$. We see that (8.22) is equivalent to

$$r\Big[1 - 2f_0(r)(r^2 - s^2)\Big]\psi_s + s\psi_r = -2rsf_0(r)\psi, \tag{8.25}$$

where

$$\psi := \phi - s\phi_s. \tag{8.26}$$

The characterize equation of the quasi-linear PDE (8.25) is

$$\frac{dr}{s} = \frac{ds}{r\left[1 - 2f_0(r)(r^2 - s^2)\right]} = \frac{d\psi}{-2rsf_0(r)\psi}. \tag{8.27}$$

It follows that $g^{-1}(r)\psi = c_1$, $g^2(r)(r^2 - s^2) = c_2$ are independent integrals of (8.27) where $g = g(r)$ is given in (8.24). Hence the solution of (8.25) is

$$\psi = g(r)f\Big((r^2 - s^2)g^2(r)\Big) \tag{8.28}$$

where f is any continuously differentiable function. Hence

$$\phi - s\phi_s = g(r)f\Big((r^2 - s^2)g^2(r)\Big). \tag{8.29}$$

It follows that every solution of (8.22) satisfies (8.29).

Conversely, suppose that (8.29) holds. Then we obtain (8.25) and (8.26). Thus ϕ satisfies (8.22). We conclude that (8.29) and (8.22) are equivalent.

Now we consider $s \in [s_0, \infty)$ where $s_0 > 0$. Put

$$\phi = s\varphi. \tag{8.30}$$

It follows that $\phi_s = \varphi + s\varphi_s$. Together with (8.28) yields

$$g(r)f\left((r^2-\sigma^2)g^2(r)\right) = s\varphi - s(\varphi + s\varphi_s) = -s^2\varphi_s.$$

Thus $\varphi = h(r) - g(r)\int_{s_0}^{s}\sigma^{-2}f\left((r^2-\sigma^2)g^2(r)\right)d\sigma$. Plugging this into (8.30) yields (8.23).

Remark 8.1.1 Similarly, we can obtain the general solution of (8.22) for $s < 0$.

We set

$$t := (r^2 - s^2)g^2(r). \tag{8.31}$$

Then (8.29) simplifies to

$$\phi - s\phi_s = g(r)f(t). \tag{8.32}$$

Differentiating (8.32) with respect to s, we obtain $-s\phi_{ss} = -2sg^3(r)f'(t)$. It follows that $(r^2 - s^2)\phi_{ss} = 2tg(r)f'(t)$. Taking this together with (8.32) yields

$$\phi - s\phi_s + (r^2 - s^2)\phi_{ss} = g(r)[f(t) + 2tf'(t)]. \tag{8.33}$$

Note that $g(r) > 0$. Considering $F(x, y) = |y|\phi\left(|x|, \frac{\langle x, y\rangle}{|y|}\right)$ where ϕ satisfies (8.23), then F is a Finsler metric if and only if the positive function ϕ satisfies (see Sect. 1.3) [82]

$$f(t) + 2tf'(t) = \frac{\phi - s\phi_s + (r^2 - s^2)\phi_{ss}}{g(r)} > 0, \quad when\ m \geq 2, \tag{8.34}$$

with the additional inequality

$$f(t) = \frac{\phi - s\phi_s}{g(r)} > 0, \quad when\ m \geq 3, \tag{8.35}$$

where $t \geq 0$. Taking $f_0(r) = \frac{\lambda b}{a+br^2}$ in (8.24) where a, b and λ are constants satisfying $a + br^2 > 0$. Thus we have

$$-\int 2rf_0(r)dr = \ln(a + br^2)^{-\lambda} + \ln c$$

where $c > 0$. Plugging this into (8.24) yields $g(r) = \frac{c}{(a+br^2)^\lambda}, \quad c > 0$. Without loss of generality we assume that $c = 1$. Then $g(r) = \frac{1}{(a+br^2)^\lambda}$. Substituting this into (8.23) we have $\phi(r, s) = s \cdot h(r) - \frac{s}{(a+br^2)^\lambda} \int_{s_0}^{s} \sigma^{-2} f\left(\frac{r^2-\sigma^2}{(a+br^2)^{2\lambda}}\right) d\sigma$. When $\lambda b = 0$, then $Q = 0$. In this case, $F = |y|\phi\left(|x|, \frac{\langle x, y\rangle}{|y|}\right)$ is projectively flat. By a straightforward computation one obtains

$$f_0'(r) = \frac{-2\lambda b^2 r}{(a + br^2)^2}.$$

Combining this with (8.20) we have $R_2 = \frac{4\lambda(\lambda-1)b^2}{(a+br^2)^2}$. It follows that the Weyl curvature of F is given by

$$\begin{aligned} W^i{}_j =& \frac{4\lambda(\lambda - 1)b^2}{(a + br^2)^2} \sum_{k,l} \left(\delta_{kl} x^i x^j + \frac{1}{m-1} \delta^i_j x^k x^l + \frac{|x|^2}{m-1} \delta^i_k \delta_{jl} \right. \\ & \left. - \delta_{jl} x^i x^k - \frac{|x|^2}{m-1} \delta^i_j \delta_{kl} - \frac{1}{m-1} \delta^i_l x^j x^k \right) y^k y^l. \end{aligned} \tag{8.36}$$

where we have used Proposition 8.1.3.

We conclude that when $\lambda b \neq 0$, then F is a Weyl metric if and only if $\lambda = 1$.

Taking $f(t) = t^n + \epsilon$ where $n \in \{1, 2, 3, \cdots\}$ and $\epsilon > 0$ we have $f(t) > 0$, $f(t) + 2tf'(t) > 0$ when $t \geq 0$. Thus f satisfies (8.34) and (8.35). Moreover,

$$\begin{aligned} \phi(r, s) =& s \cdot h(r) - \frac{s}{(a + br^2)^\lambda} \int_{s_0}^{s} \sigma^{-2} \left[\frac{(r^2 - \sigma^2)^n}{(a + br^2)^{2\lambda n}} + \epsilon \right] d\sigma \\ =& s \cdot h_1(r) + \frac{\epsilon}{(a + br^2)^\lambda} - \frac{s}{(a + br^2)^{(2n+1)\lambda}} \int_{s_0}^{s} \sigma^{-2}(r^2 - \sigma^2)^n d\sigma. \end{aligned} \tag{8.37}$$

We require Lemma 6.3 in Chap. 6. By using (6.44) and (8.37), we have

$$\begin{aligned}
\phi(r,s) &= s \cdot h_2(r) - \frac{s}{(a+br^2)^{(2n+1)\lambda}} \left\{ \frac{(r^2-s^2)^n}{(2n-1)s} \right. \\
&\quad + \frac{n!}{(2n-1)!!s} \left[\sum_{i=2}^{n} \frac{(2n-2i-1)!!}{(n-i+1)!} (2r^2)^{i-1} (r^2-s^2)^{n-i+1} \right. \\
&\quad \left. \left. - (2r^2)^n \right] + C(r) \right\} + \frac{\epsilon}{(a+br^2)^\lambda} \\
&= s \cdot h_3(r) - \frac{(r^2-s^2)^n}{(2n-1)(a+br^2)^{(2n+1)\lambda}} \\
&\quad + \frac{n!}{(2n-1)!!} \frac{(2r^2)^n}{(a+br^2)^{(2n+1)\lambda}} + \frac{\epsilon}{(a+br^2)^\lambda} \\
&\quad - \frac{1}{(a+br^2)^{(2n+1)\lambda}} \sum_{i=2}^{n} \frac{n!(2n-2i-1)!!}{(2n-1)!!(n-i+1)!} (2r^2)^{i-1} (r^2-s^2)^{n-i+1}.
\end{aligned} \tag{8.38}$$

Together with (8.36) we have the following:

Theorem 8.1.1 *Let $\phi(r,s)$ be a function defined by*

$$\begin{aligned}
\phi(r,s) &= s \cdot h(r) + \frac{n!}{(2n-1)!!} \frac{(2r^2)^n}{(a+br^2)^{(2n+1)\lambda}} - \frac{(r^2-s^2)^n}{(2n-1)(a+br^2)^{(2n+1)\lambda}} \\
&\quad - \frac{1}{(a+br^2)^{(2n+1)\lambda}} \sum_{i=2}^{n} \frac{n!(2n-2i-1)!!}{(2n-1)!!(n-i+1)!} (2r^2)^{i-1} (r^2-s^2)^{n-i+1} \\
&\quad + \frac{\epsilon}{(a+br^2)^\lambda}
\end{aligned}$$

where $n \in \{1,2,3,\cdots\}$, ϵ, a, b and λ are constants satisfying $\epsilon > 0$ and $a+br^2 > 0$ and $h(r)$ is a differential function. Then on $\mathbb{B}^m(r_\mu)$ the following Finsler metric

$$F(x,y) = |y|\phi\left(|x|, \frac{\langle x, y\rangle}{|y|}\right)$$

is W-quadratic. The Weyl curvature of F is given by

$$\begin{aligned}
W^i_j = -\frac{4\lambda(\lambda-1)b^2}{(a+br^2)^2} &\left[\frac{1}{m-1} \left(|x|^2|y|^2 - \langle x, y\rangle^2 \right) \delta^i_j - |y|^2 x^i x^j \right. \\
&\left. + \langle x, y\rangle x^i y^j + \frac{\langle x, y\rangle}{m-1} x^j y^i - \frac{|x|^2}{m-1} y^i y^j \right].
\end{aligned}$$

Furthermore, F is a Weyl metric if and only if $\lambda \in \{0, 1\}$ or $b = 0$; F is projectively flat if and only if $\lambda b = 0$.

When $a = 1$, $\lambda(\lambda - 1)b = 0$ and $\epsilon = 0$ our Weyl metrics have been constructed in Chap. 6.

8.1.4 Non-trivial Finsler Metrics of Quadratic Weyl Curvature

A Finsler metric is said to be *R-quadratic* if its Riemannian curvature R_y are quadratic in $y \in T_xM$ [36, 44, 65]. For example, all Berwald metrics are R-quadratic.

By the definition of $W^i{}_j$ (see (8.1) and (8.2)), if $R^i{}_j$ are quadratic in y, then $W^i{}_j$ are quadratic in y. Namely, every R-quadratic Finsler metric must be W-quadratic. It seems that W-quadratic Finsler metrics form a broader class than R-quadratic Finsler metrics. However, so far people do not have any explicit examples of W-quadratic Finsler metrics which are non-trivial in the sense that these metrics are not R-quadratic. In this subsection we find many non-trivial W-quadratic Finsler metrics. Precisely we prove the following:

Theorem 8.1.2 *On $\mathbb{B}^n(r_\mu)$, the following Finsler metric*

$$F = \sqrt{f(|x|)|y|^2 + \tau^2 f^2(|x|)\langle x, y\rangle^2} + \tau f(|x|)\langle x, y\rangle \tag{8.39}$$

is W-quadratic where $r = |x|$, $f_r = \frac{\partial f}{\partial r}$, $f : [0, r_\mu) \to \mathbb{R}^+$ is an any positive differentiable function and τ is a constant. Furthermore F is not R-quadratic whenever

$$\tau \frac{2f(|x|) + |x| f_r(|x|)}{1 + \tau^2|x|^2 f(|x|)} \neq constant.$$

Let us take a look at the special case: when $f(r) = \frac{\epsilon}{1+\zeta r^2}$,

$$F(x, y) = \frac{\sqrt{\tau^2\langle x, y\rangle^2 + \epsilon|y|^2(1 + \zeta|x|^2)}}{1 + \zeta|x|^2} + \frac{\tau\langle x, y\rangle}{1 + \zeta|x|^2},$$

where ζ is a constant and ϵ is a positive constant. F is W-quadratic. However F is not R-quadratic unless $\epsilon\zeta + \tau^2 = 0$.

Proof (8.39) can be expressed in the following form $F = \alpha\phi\left(b^2, \frac{\beta}{\alpha}\right)$ where $\alpha = |y|$, $\beta = \langle x, y\rangle$, $b^2 = |x|^2$, $\phi = \sqrt{f(\sqrt{b^2}) + \tau^2 f^2(\sqrt{b^2})s^2} + \tau f(\sqrt{b^2})s$. Then ${}^\alpha Ric = 0$, $b_{i|j} = a_{ij}$. In [9], Chen-Shen show that a Finsler metric is of isotropic Berwald curvature if and only if it is a Douglas metric (i.e. $\mathscr{D} = 0$) with isotropic mean Berwald curvature where $\mathscr{D}$ denotes the Douglas curvature. Combining this with Theorem 3.3 we get F satisfies the conditions of Theorem 4.2 in [46]. It follows that F is W-quadratic.

We know that R-quadratic Randers metrics must have constant S-curvature [36]. It follows that F is not R-quadratic whenever $\tau\frac{2f(|x|)+|x|f_r(|x|)}{1+\tau^2|x|^2f(|x|)} \neq constant.$

We know that all R-quadratic Finsler metrics have vanishing H-curvature. Therefore an interesting problem is to ask if all W-quadratic Finsler metrics have vanishing H-curvature.

8.2 Projectively Related Spherically Symmetric Metrics

In this section we find an equation that characterizes projectively related spherically symmetric metric. More precisely, we show the following:

Theorem 8.2.1 *Let $F = |y|\phi(r, s)$ and $\tilde{F} = |y|\tilde{\phi}(r, s)$ be two spherically symmetric Finsler metrics on $\mathbb{B}^n(r_\mu)$, where $r := |x|$, $s := \frac{\langle x, y\rangle}{|y|}$. Then $\tilde{F}$ is pointwise projectively related to F if and only if ϕ and $\tilde{\phi}$ satisfy*

$$\frac{r\phi_{ss} - \phi_r + s\phi_{rs}}{\phi - s\phi_s + (r^2 - s^2)\phi_{ss}} = \frac{r\tilde{\phi}_{ss} - \tilde{\phi}_r + s\tilde{\phi}_{rs}}{\tilde{\phi} - s\tilde{\phi}_s + (r^2 - s^2)\tilde{\phi}_{ss}}.$$

We know that every Weyl metric must be W-quadratic. In [36], Li-Shen find equations that characterize W-quadratic Randers metrics. It seems that W-quadratic Randers metrics form a broader class than Weyl Randers metrics although there is no example supporting this. As an important application of Theorem 8.2.1, in this section we find a lot of Randers metrics of quadratic Weyl curvature which are non-trivial in the sense that they are not of Weyl type.

Theorem 8.2.2 *Let $\tilde{F}(x, y)$ be a Finsler metric on $B^n(r_\mu)$ defined by*

$$\tilde{F}(x, y) = \frac{\sqrt{\kappa^2\langle x, y\rangle^2 + \varepsilon|y|^2(1 + \zeta|x|^2)}}{1 + \zeta|x|^2} + \frac{\kappa\langle x, y\rangle}{1 + \zeta|x|^2}.$$

where κ, ζ and ε are arbitrary constants such that $\varepsilon > 0$; $\mu = 1/\sqrt{-\zeta}$ if $\zeta < 0$ and $\mu = +\infty$ if $\zeta \geq 0$. Then the Weyl curvature of $\tilde{F}$ is given by

$$W^i{}_j = -\frac{(\zeta\varepsilon + \kappa^2)^2}{\left[(\zeta\varepsilon + \kappa^2)r^2 + \varepsilon\right]^2}\Bigg[\frac{1}{n-1}\left(|x|^2|y|^2 - \langle x, y\rangle^2\right)\delta^i_j - |y|^2x^ix^j$$
$$+ \langle x, y\rangle x^iy^j + \frac{\langle x, y\rangle}{n-1}x^jy^i - \frac{|x|^2}{n-1}y^iy^j\Bigg], \tag{8.40}$$

with $r = |x|$, *therefore* $\tilde{F}$ *is a non-trivial* W*-quadratic Randers metric when* $\zeta\varepsilon + \kappa^2 \neq 0$.

For a proof of Theorem 8.2.2, see Sect. 8.2.3 below. In particular, we show that Chern-Shen's Randers metrics are non-trivial W-quadratic Finsler metrics with isotropic S-curvature, see Corollary 8.2.5 below.

8.2.1 Reducible Differential Equation

Let $F = |y|\phi\left(|x|, \frac{\langle x,y\rangle}{|y|}\right)$ be a spherically symmetric Finsler metric on $\mathbb{B}^n(r_\mu)$. Then its geodesic coefficients are given in (3.1), where P and Q satisfy (3.3) and (3.2) respectively.

Let $\tilde{F} = |y|\tilde{\phi}\left(|x|, \frac{\langle x,y\rangle}{|y|}\right)$ be another spherically symmetric Finsler metric on $\mathbb{B}^n(r_\mu)$. Then we have

$$\tilde{G}^i = |y|\tilde{P}y^i + |y|^2\tilde{Q}x^i \tag{8.41}$$

where we denote the corresponding objects with respect to $\tilde{F}$ by adding a tilde $\tilde{}$.

Proof of Theorem 8.2.1 Assume that $\tilde{F}$ is pointwise projectively related to F. Then [11, 18, 64, 67, 83]

$$\tilde{G}^i = G^i + Ry^i \tag{8.42}$$

where R is positively homogeneous of degree one. Plugging (8.41) and (3.17) into (8.42) yields $|y|^2(\tilde{Q} - Q)x^i + \left[|y|(\tilde{P} - P) - R\right]y^i = 0$. It follows that

$$\tilde{Q} = Q, \qquad |y|\tilde{P} = |y|P + R. \tag{8.43}$$

Conversely, suppose that the first equation of (8.43) holds for two spherically symmetric Finsler metrics $F = |y|\phi\left(|x|, \frac{\langle x,y\rangle}{|y|}\right)$ and $\tilde{F} = |y|\tilde{\phi}\left(|x|, \frac{\langle x,y\rangle}{|y|}\right)$. By using (3.1) and (8.41) we have

$$\tilde{G}^i - G^i = |y|\tilde{P}y^i + |y|^2\tilde{Q}x^i - |y|Py^i - |y|^2Qx^i = Ry^i,$$

where $R := |y|(\tilde{P} - P)$. Together with (3.3), we obtain that R is positively homogeneous of degree one. According to Theorem 2.1 in [18] or (2.2) in [64], $\tilde{F}$ must be pointwise projectively related to F. The above arguments and (3.2) complete the proof of Theorem 8.2.1.

As a consequence of Theorem 8.2.1, by taking the standard Euclidean metric $\tilde{F}$, we obtain the following result obtained by Huang and the second author (see Theorem 5.1):

Corollary 8.2.1 *Let $F = |y|\phi\left(|x|, \frac{\langle x, y\rangle}{|y|}\right)$ be a spherically symmetric Finsler metric on $\mathbb{B}^n(r_\mu)$. Then F is projectively flat if and only if ϕ satisfies $r\phi_{ss} - \phi_r + s\phi_{rs} = 0$.*

Corollary 8.2.2 *The pointwise projective relatedness for two spherically symmetric Finsler metrics is independent of the dimension of the base space.*

Proposition 8.2.1 *Let $F_1 = |y|\phi_1\left(|x|, \frac{\langle x, y\rangle}{|y|}\right)$ and $F_2 = |y|\phi_2\left(|x|, \frac{\langle x, y\rangle}{|y|}\right)$ be two spherically symmetric Finsler metrics on $\mathbb{B}^n(r_\mu)$ with*

$$\begin{aligned}\phi_j(r,s) = s\cdot h_j(r) &+ \frac{m_j!}{(2m_j-1)!!}\frac{(2r^2)^{m_j}}{(a_j+b_jr^2)^{(2m_j+1)\lambda_j}} - \frac{(r^2-s^2)^{m_j}}{(2m_j-1)(a_j+b_jr^2)^{(2m_j+1)\lambda_j}}\\ &- \frac{1}{(a_j+b_jr^2)^{(2m_j+1)\lambda_j}}\sum_{i=2}^{m_j}\frac{m_j!(2m_j-2i-1)!!}{(2m_j-1)!!(m_j-i+1)!}(2r^2)^{i-1}(r^2-s^2)^{m_j-i+1}\\ &+ \frac{\epsilon_j}{(a_j+b_jr^2)^{\lambda_j}}\end{aligned} \tag{8.44}$$

where $m_j \in \{1, 2, 3, \cdots\}$, ϵ_j, a_j, b_j and λ_j are constants satisfying $\epsilon_j > 0$ and $a_j + b_jr^2 > 0$ and h_j are differentiable functions. Then F_1 is pointwise projectively related to F_2 if and only if a_j, b_j and λ_j $(j = 1, 2)$ satisfy

$$b_1b_2(\lambda_1-\lambda_2) = 0, \qquad \lambda_1b_1a_2 = \lambda_2b_2a_1. \tag{8.45}$$

Furthermore, F_1 and F_2 are projectively flat if $b_1b_2 = 0$ or $b_1b_2 \neq 0$, $\lambda_1 = \lambda_2 = 0$; F_1 and F_2 are of scalar curvature if $b_1b_2 = 0$ or $b_1b_2 \neq 0$, λ_1 or $\lambda_2 \in \{0, 1\}$; F_1 and F_2 are non-trivial W-quadratic metrics if $\lambda_1(\lambda_1-1)b_1 \neq 0$ or $\lambda_2(\lambda_2-1)b_2 \neq 0$.

Proof According to Theorem 8.2.1, F_1 is pointwise projectively related to F_2 if and only if

$$Q_1 = Q_2 \tag{8.46}$$

where

$$Q_j = \frac{1}{2r}\frac{r(\phi_j)_{ss} - (\phi_j)_r + s(\phi_j)_{rs}}{(\phi_j) - s(\phi_j)_s + (r^2-s^2)(\phi_j)_{ss}}, \qquad j = 1, 2. \tag{8.47}$$

On the other hand, (8.44) gives solutions of (8.47) with

$$Q_j := \frac{\lambda_jb_j}{a_j+b_jr^2}, \qquad j = 1, 2 \tag{8.48}$$

where a_j, b_j and λ_j are constants satisfying $a_j + b_j r^2 > 0$ [10]. Thus F_1 is pointwise projectively related to F_2 if and only if

$$\frac{\lambda_1 b_1}{a_1 + b_1 r^2} = \frac{\lambda_2 b_2}{a_2 + b_2 r^2}. \tag{8.49}$$

It is easy to see that (8.49) holds if and only if $b_1 b_2(\lambda_1 - \lambda_2) r^2 + (\lambda_1 b_1 a_2 - \lambda_2 b_2 a_1) = 0$. Together with the third equation of (3.2) we have (8.45). "Furthermore ..." is an immediate consequence of Theorem 8.1.1.

Theorem 8.2.3 *Let* $F_1 = |y|\phi_1\left(|x|, \frac{\langle x,y\rangle}{|y|}\right)$ *and* $F_2 = |y|\phi_2\left(|x|, \frac{\langle x,y\rangle}{|y|}\right)$ *be two spherically symmetric Finsler metrics on* $\mathbb{B}^n(r_\mu)$ *with*

$$\begin{aligned}\phi_j(r,s) = s \cdot h_j(r) + &\frac{m_j!}{(2m_j-1)!!}\frac{(2r^2)^{m_j}}{(a_j+b_jr^2)^{2m_j+1}} - \frac{(r^2-s^2)^{m_j}}{(2m_j-1)(a_j+b_jr^2)^{2m_j+1}}\\ &-\frac{1}{(a_j+b_jr^2)^{2m_j+1}}\sum_{i=2}^{m_j}\frac{m_j!(2m_j-2i-1)!!}{(2m_j-1)!!(m_j-i+1)!}(2r^2)^{i-1}(r^2-s^2)^{m_j-i+1}\\ &+\frac{\epsilon_j}{a_j+b_jr^2}\end{aligned}$$

where $m_j \in \{1, 2, 3, \cdots\}$, ϵ_j, a_j *and* b_j *are constants satisfying* $\epsilon_j > 0$, $a_j + b_j r^2 > 0$ *and* h_j *are differentiable functions. Then* F_1 *is pointwise projectively related to* F_2 *if and only if* a_1, a_2, b_1 *and* b_2 *satisfy* $b_1 a_2 = b_2 a_1$.

Proof The result follows from Proposition 8.2.1 by taking $\lambda_1 = \lambda_2 = 1$.

Theorem 8.2.3 tells us that the pointwise projective relatedness of two Huang-Mo's spherically symmetric Weyl metrics is independent of the functions h_j, the natural numbers m_j and the constants ϵ_j, $j = 1, 2$.

8.2.2 *Projectively Related Weyl Quadratic Metrics*

Given a Finsler metric on a manifold M, a natural problem is to determine all Finsler metrics which are pointwise projectively related to the given metric [67].

In this subsection, we study the following problem: given a Weyl quadratic spherically symmetric Finsler metric, describe all spherically symmetric Finsler metrics which are pointwise projectively related to the given one.

Proposition 8.2.2 *Let ϕ be a function defined by*

$$\phi(r,s) = s\cdot h(r) + \frac{m!}{(2m-1)!!}\frac{(2r^2)^m}{(a+br^2)^{(2m+1)\lambda}} - \frac{(r^2-s^2)^m}{(2m-1)(a+br^2)^{(2m+1)\lambda}}$$
$$-\frac{1}{(a+br^2)^{(2m+1)\lambda}}\sum_{i=2}^{m}\frac{m!(2m-2i-1)!!}{(2m-1)!!(m-i+1)!}(2r^2)^{i-1}(r^2-s^2)^{m-i+1}$$
$$+\frac{\epsilon}{(a+br^2)^{\lambda}} \tag{8.50}$$

where $m \in \{1, 2, 3, \cdots\}$; ϵ, a, b and λ are constants such that $\epsilon > 0$ and $a+br^2 > 0$ and h are differentiable functions. Then any spherically symmetric Finsler metric which is pointwise projectively related to $F = |y|\phi\left(|x|, \frac{\langle x,y\rangle}{|y|}\right)$ is given by

$$\tilde{F} = |y|\tilde{\phi}\left(|x|, \frac{\langle x,\ y\rangle}{|y|}\right)$$

where

$$\tilde{\phi}(r,s) = s\left(f(r) - \int \frac{\eta(\varphi(r,s))}{s^2\sqrt{r^2-s^2}}ds\right), \tag{8.51}$$

with φ given by

$$\varphi(r,s) = -\frac{r^2-s^2}{(a+br^2)^{2\lambda}}; \tag{8.52}$$

f and η are arbitrary differentiable real functions of r and φ respectively, and

$$\frac{-\sqrt{r^2-s^2}}{s}\frac{\partial\eta}{\partial s} > 0,\ \text{when}\ n \geq 2,$$

with the additional inequality

$$\frac{\eta}{\sqrt{r^2-s^2}} > 0,\ \text{when}\ n \geq 3,$$

where $r^2 - s^2 > 0$ and $s \neq 0$.

Proof According to the proof of Theorem 8.1.1, we have

$$\frac{1}{2r}\frac{r\phi_{ss} - \phi_r + s\phi_{rs}}{\phi - s\phi_s + (r^2-s^2)\phi_{ss}} = Q = \frac{\lambda b}{a+br^2} \tag{8.53}$$

where we have used the first equation of (3.2). Together with Theorem 8.2.1 we obtain

$$\left[(r^2-s^2)\frac{2\lambda b}{a+br^2}-1\right]r\tilde{\phi}_{ss}+\tilde{\phi}_r-s\tilde{\phi}_{rs}+\frac{2\lambda br}{a+br^2}(\tilde{\phi}-s\tilde{\phi}_s)=0. \tag{8.54}$$

By Lemma 4.1 in [54] and (8.53), F has vanishing Douglas curvature. Then Proposition 8.2.2 follows from Theorem 1.2 in [54].

As a consequence of Proposition 8.2.2, by taking $\lambda = 1$, we obtain the following:

Theorem 8.2.4 *Let ϕ be a function defined by*

$$\begin{aligned}\phi(r,s) &= s\cdot h(r)+\frac{m!}{(2m-1)!!}\frac{(2r^2)^m}{(a+br^2)^{2m+1}}-\frac{(r^2-s^2)^m}{(2m-1)(a+br^2)^{2m+1}}\\ &\quad -\frac{1}{(a+br^2)^{2m+1}}\sum_{i=2}^{m}\frac{m!(2m-2i-1)!!}{(2m-1)!!(m-i+1)!}(2r^2)^{i-1}(r^2-s^2)^{m-i+1}\\ &\quad +\frac{\epsilon}{a+br^2}\end{aligned}$$

where $m\in\{1,2,3,\cdots\}$, ϵ, a and b are constants such that $\epsilon>0$ and $a+br^2>0$ and h is a differentiable function. Then any spherically symmetric Finsler metric which is pointwise projectively related to $F=|y|\phi\left(|x|,\frac{\langle x,y\rangle}{|y|}\right)$ is given by

$$\tilde{F}=|y|\tilde{\phi}\left(|x|,\frac{\langle x,y\rangle}{|y|}\right)$$

where $\tilde{\phi}(r,s)=s\left(f(r)-\int\frac{\eta(\varphi(r,s))}{s^2\sqrt{r^2-s^2}}ds\right)$, with φ given by

$$\varphi(r,s)=-\frac{r^2-s^2}{(a+br^2)^2};$$

f and η are arbitrary differentiable real functions of r and φ respectively, and

$$\frac{-\sqrt{r^2-s^2}}{s}\frac{\partial\eta}{\partial s}>0,\ \text{when } n\geq 2,$$

with the additional inequality

$$\frac{\eta}{\sqrt{r^2-s^2}}>0,\ \text{when } n\geq 3,$$

where $r^2-s^2>0$ and $s\neq 0$.

8.2.3 Non-trivial W-Quadratic Randers Metrics

In this subsection, we are going to find Randers metrics of quadratic Weyl curvature which are non-trivial in the sense that they are not of Weyl type. Recall that a Finsler metric is called a *Weyl metric* if it has vanishing Weyl curvature [28, 38, 59] and Finsler metric is said to be *W-quadratic* if it has quadratic Weyl curvature [13, 36]. According to M. Matsumoto's result, a Finsler metric is of Weyl type if and only if it is of scalar curvature.

As a consequence of Proposition 8.2.2, for $\lambda = \frac{1}{2}$, $b = \zeta\delta + \kappa^2$, $a = \delta$ and hence $\varphi(r, s) = \frac{-(r^2 - s^2)}{(\zeta\delta + \kappa^2)r^2 + \delta}$, with the choice $\eta(\varphi) = \delta\sqrt{-\left(\frac{1}{\varphi} + \kappa^2\right)^{-1}}$, we get the following result:

Corollary 8.2.3 *Let $\phi(r, s)$ be a function defined by*

$$\begin{aligned}\phi(r, s) = s \cdot h(r) + \frac{m!}{(2m-1)!!}\frac{(2r^2)^m}{\left[(\zeta\delta + \kappa^2)r^2 + \delta\right]^{\frac{2m+1}{2}}} - \frac{(r^2 - s^2)^m}{(2m-1)\left[(\zeta\delta + \kappa^2)r^2 + \delta\right]^{\frac{2m+1}{2}}}\\ - \frac{1}{\left[(\zeta\delta + \kappa^2)r^2 + \delta\right]^{\frac{2m+1}{2}}}\sum_{i=2}^{m}\frac{m!(2m-2i-1)!!}{(2m-1)!!(m-i+1)!}(2r^2)^{i-1}(r^2 - s^2)^{m-i+1}\\ + \frac{\epsilon}{\left[(\zeta\delta + \kappa^2)r^2 + \delta\right]^{\frac{1}{2}}}\end{aligned} \tag{8.55}$$

where $m \in \{1, 2, 3, \cdots\}$, δ, ε, ζ and κ are constants such that $\delta > 0$ and $(\zeta\delta + \kappa^2)r^2 + \delta > 0$ and h is a differentiable function. Then the spherical symmetric Finsler metric

$$F = |y|\phi\left(|x|, \frac{\langle x, y\rangle}{|y|}\right) \tag{8.56}$$

is pointwise projectively related to $\tilde{F}(x, y) = |y|\tilde{\phi}\left(|x|, \frac{\langle x, y\rangle}{|y|}\right)$ where $\tilde{\phi}(r, s) = sf(r) + \frac{\sqrt{\zeta\delta r^2 + \kappa^2 s^2 + \delta}}{\zeta r^2 + 1}$ with any real function f such that $\tilde{\phi}(r, s)$ is positive.

In particular, when $f(r) = \frac{\kappa}{1 + \zeta r^2}$, we have the following:

Corollary 8.2.4 *Let ϕ be a function defined by* (8.55). *Then the spherical symmetric metric*

$$F = |y|\phi\left(|x|, \frac{\langle x, y\rangle}{|y|}\right)$$

is pointwise projectively related to the Randers metric

$$\tilde{F}(x, y) = \frac{\sqrt{\kappa^2\langle x, y\rangle^2 + \delta|y|^2(1+\zeta|x|^2)}}{1+\zeta|x|^2} + \frac{\kappa\langle x, y\rangle}{1+\zeta|x|^2}. \tag{8.57}$$

Proof of Theorem 8.2.2. According to Corollary 8.2.4, $\tilde{F}(x, y)$ is pointwise projectively related to $F = |y|\phi\left(|x|, \frac{\langle x, y\rangle}{|y|}\right)$ where ϕ is defined by (8.55). Since the Weyl curvature is a projective invariant, Theorem 8.1.1 gives (8.40).

When $0 < \zeta = \varepsilon$ and $\kappa^2 = 1 - \varepsilon^2$ we get the following:

Corollary 8.2.5 *Let F be a Finsler metric defined by*

$$F(x, y) := \frac{\sqrt{(1-\varepsilon^2)\langle x, y\rangle^2 + \varepsilon|y|^2(1+\varepsilon|x|^2)}}{1+\varepsilon|x|^2} + \frac{\sqrt{1-\varepsilon^2}\langle x, y\rangle}{1+\varepsilon|x|^2}$$

with $0 < \varepsilon \leq 1$. Then F is a non-trivial W-quadratic Randers metric of isotropic S-curvature. Moreover, the Weyl curvature of F is given by

$$W^i{}_j = -\frac{1}{[\varepsilon+|x|^2]^2}\Bigg[\frac{1}{n-1}\left(|x|^2|y|^2 - \langle x, y\rangle^2\right)\delta^i_j - |y|^2x^ix^j$$
$$+\langle x, y\rangle x^iy^j + \frac{\langle x, y\rangle}{n-1}x^jy^i - \frac{|x|^2}{n-1}y^iy^j\Bigg].$$

References

1. J.C. Álvarez, *Hilbert's Fourth Problem in Two Dimensions*. MASS selecta, 165–183 (American Mathematical Society, Porvidence, 2003)
2. H. Alzer, Error function inequalities. Adv. Comput. Math. **33**, 349–379 (2010)
3. S.-I. Amari, H. Nagaoka, *Methods of Information Geometry*. AMS Translations of Mathematical Monographs (Oxford University Press, Oxford, 2000)
4. D. Bao, Z. Shen, Finsler metrics of constant positive curvature on the Lie group S^3. J. Lond. Math. Soc. **66**, 453–467 (2002)
5. L. Berwald, Untersuchung der Krümmung allgemeiner metrischer Räume auf Grund des in ihnen herrschenden Parallelismus. Math. Z. **25**, 40–73 (1926)
6. L. Berwald, *Ü*ber die n-dimensionalen Geometrien konstanter Kr*ü*mmung, in denen die Geraden die k*ü*rzesten sind. Math. Z. **30**, 449–469 (1929)
7. R. Bryant, Projectively flat Finsler 2-spheres of constant curvature. Selecta Math. **3**, 161–203 (1997)
8. R. Bryant, Some remarks on Finsler manifolds with constant flag curvature. Spec. Issue S. S. Chern. Houst. J. Math. **28**, 221–262 (2002)
9. X. Cheng, Z. Shen, On Douglas metrics. Publ. Math. Debr. **66**, 503–512 (2007)
10. X. Cheng, Z. Shen, Randers metrics with special curvature properties. Osaka J. Math. **40**, 87–101 (2003)
11. G. Chen, X. Cheng, A class of Finsler metrics projectively related to a Randers metric. Publ. Math. Debr. **81**, 351–363 (2012)
12. X. Chen, X. Mo, Z. Shen, On the flag curvature of Finsler metrics of scalar curvature. J. Lond. Math. Soc. **68**, 762–780 (2003)
13. X. Cheng, Z. Shen, *Finsler Geometry, An Approach via Randers Space* (Science Press, Beijing, 2012)
14. X. Cheng, Z. Shen, Y. Zhou, On locally dually flat Finsler metrics. Int. J. Math. **21**, 1531–1543 (2010)
15. X. Cheng, Y. Tian, Locally dually flat Finsler metrics with special curvature properties. Differ. Geom. Appl. **29**, 98–106 (2011)
16. S.S. Chern, Finsler geometry is just Riemannian geometry without the quadratic restriction on its metrics. Not. Am. Math. Soc. **43**, 959–963 (1996)
17. S.S. Chern, Z. Shen, *Riemann-Finsler Geometry*. Nankai Tracts in Mathematics, vol. 6 (World Scientific, Hackensack, 2005), x+192 pp
18. N. Cui, Y. Shen, Projective change between two classes of$(\alpha,\ \beta)$-metrics. Differ. Geom. Appl. **27**, 566–573 (2009)

E. Guo, X. Mo, *The Geometry of Spherically Symmetric Finsler Manifolds*, SpringerBriefs in Mathematics, https://doi.org/10.1007/978-981-13-1598-5

19. T. Ding, C. Li, *Lecture on Ordinary Differential Equations*, 2nd edn. (Higher Education Press, Beijing, 2004)
20. J. Douglas, The general geometry of paths. Ann. Math. **29**, 143–168 (1927–1928)
21. E. Guo, H. Liu, X. Mo, On spherically symmetric Finsler metrics with isotropic Berwald curvature. Int. J. Geom. Methods Mod. Phys. **10**, 1350054 (2013), 13 pp
22. P. Funk, Über Geometrien bei denen die Geraden die Kürzesten sind. Math. Ann. **101**, 226–237 (1929)
23. G. Hamel, *Ü*ber die Geometrien in denen die Geraden die K*ü*rzesten sind. Math. Ann. **57**, 231–264 (1903)
24. L. Huang, Navigation problem on Finsler manifold and its applications, Ph.D. thesis (2008)
25. L. Huang, X. Mo, Projectively flat Finsler metrics with orthogonal invariance. Ann. Polon. Math. **107**, 259–270 (2013)
26. L. Huang, X. Mo, On geodesics of Finsler metrics via navigation problem. Proc. Am. Math. Soc. **139**, 3015–3024 (2011)
27. L. Huang, X. Mo, A new class of projectively flat Finsler metrics in terms of hypergeometric functions. Publ. Math. Debr. **81**, 421–434 (2012)
28. L. Huang, X. Mo, On spherically symmetric Finsler metrics of scalar curvature. J. Geom. Phys. **62**, 2279–2287 (2012)
29. L. Huang, X. Mo, On some explicit constructions of dually flat Finsler metrics. J. Math. Anal. Appl. **405**, 565–573 (2013)
30. L. Huang, X. Mo, On some dually flat Finsler metrics with orthogonal invariance. Nonlinear Anal. **108**, 214–222 (2014)
31. S.-M. Jung, Hyers-Ulam stability for Gegenbauer differential equations. Electron. J. Differ. Equ. **156**, 8 (2013)
32. A. Katok, Ergodic perturbations of degenerate integrable Hamiltonian systems. Izv. Akad. Nauk SSSR Ser. Mat. **37**, 539–576 (1973)
33. B. Li, On dually flat Finsler metrics. Differ. Geom. Appl. **31**, 718–724 (2013)
34. B. Li, On the classification of projectively flat Finsler metrics with constant flag curvature. Adv. Math. **257**, 266–284 (2014)
35. B. Li, Z. Shen, On a class of projectively flat Finsler metrics with constant flag curvature. Int. J. Math. **18**, 749–760 (2007)
36. B. Li, Z. Shen, On Randers metrics of quadratic Riemann curvature. Int. J. Math. **20**, 369–376 (2009)
37. B. Li, Z. Shen, Projectively flat fourth root Finsler metrics. Can. Math. Bull. **55**, 138–145 (2012)
38. H. Liu, X. Mo, Examples of Finsler metrics with special curvature properties. Math. Nachr. **288**, 1527–1537 (2015)
39. H. Liu, X. Mo, The explicit construction of all dually flat Randers metrics. Int. J. Math. **28**, 1750058 (2017), 12 pp
40. J.L. Lopez, S.E. Perez, The role of the error function in a singularly perturbed convection-diffusion problem in a rectangle with corner singularities. Proc. R. Soc. Edinb. Sect. A **137**, 93–109 (2007)
41. J.C. Álvarez Paiva, *Hilbert's Fourth Problem in Two Dimensions*. MASS Selecta, 165–183 (American Mathematical Society, Providence, 2003)
42. M. Matsumoto, Projective changes of Finsler metrics and projectively flat Finsler spaces. Tensor N. S. **34**, 303–315 (1980)
43. P.J. McCarthy, S.F. Rutz, The general four-dimensional spherically symmetric Finsler space. Gen. Relativ. Gravit. **25**, 589–602 (1993)
44. X. Mo, On the non-Riemannian quantity H of a Finsler metric. Differ. Geom. Appl. **27**, 7–14 (2009)
45. X. Mo, On some projectively flat Finsler metrics in terms of hypergeometric functions. Isr. J. Math. **184**, 59–78 (2011)
46. X. Mo, Finsler metrics with special Riemannian curvature properties. Differ. Geom. Appl. **48**, 61–71 (2016)

47. X. Mo, Finsler metrics with constant (or scalar) flag curvature. Proc. Indian Acad. Sci. Math. Sci. **122**, 411–427 (2012)
48. X. Mo, Flag curvature tensor on a closed Finsler surface. Results Math. **36**, 149–159 (1999)
49. X. Mo, On the flag curvature of a Finsler space with constant S-curvature. Houst. J. Math. **31**, 131–144 (2005)
50. X. Mo, *An Introduction to Finsler Geometry* (World Scientific, Hackensack, 2006)
51. X. Mo, L. Huang, On curvature decreasing property of a class of navigation problems. Publ. Math. Debr. **71**, 141–163 (2007)
52. X. Mo, Z. Shen, On negatively curved Finsler manifolds of scalar curvature. Can. Math. Bull. **48**, 112–120 (2005)
53. X. Mo, Z. Shen, C.Yang, Some constructions of projectively flat Finsler metrics. Sci. China Ser. A **49**, 703–714 (2006)
54. X. Mo, N.M. Solorzano, K. Tenenblat, On spherically symmetric Finsler metrics with vanishing Douglas curvature. Differ. Geom. Appl. **31**, 746–758 (2013)
55. X. Mo, C. Yu, On some explicit constructions of Finsler metrics with scalar flag curvature. Can. J. Math. **62**, 1325–1339 (2010)
56. X. Mo, L. Zhou, The curvatures of spherically symmetric Finsler metrics in R^n. arXiv:1202.4543
57. X. Mo, L. Zhou, H. Zhu, On a class of Finsler metrics of constant curvature. Houst. J. Math. **43**, 829–846 (2017)
58. X. Mo, H. Zhu, On a class of projectively flat Finsler metrics of negative constant flag curvature. Int. J. Math. **23**, 1250084 (2012), 14 pp
59. B. Najafi, Z. Shen, A. Tayebi, On a projective class of Finsler metrics. Publ. Math. Debr. **70**, 211–219 (2007)
60. Y. Pinchover, J. Rubinstein, *An Introduction to Partial Differential Equations* (Cambridge University Press, Cambridge, 2005), xii+371 pp
61. A.V. Pogorelov, *Hilbert's Fourth Problem* (Trans.: by R.A. Silverman). Scripta in Mathematics (V. H. Winston and Sons, Washington, DC); A Halsted Press book, John Wiley and Sons, New York/Toroto, Qut.-London, 1979, p. vi+97
62. G. Randers, On an asymmetric metric in the four-space of general relativity. Phys. Rev. **59**, 195–199 (1941)
63. S.F. Rutz, Symmetry in Finsler spaces. Contemp. Math. **196**, 289–300 (1996)
64. Y. Shen, Y. Yu, On projectively related Randers metrics. Int. J. Math. **19**, 503–520 (2008)
65. Z. Shen, On R-quadratic Finsler spaces. Publ. Math. Debr. **58**, 263–274 (2001)
66. Z. Shen, *Lectures on Finsler Geometry* (World Scientific, Singapore, 2001), xiv+307 pp
67. Z. Shen, On projectively related Einstein metrics in Riemann-Finsler geometry. Math. Ann. **320**, 625–647 (2001)
68. Z. Shen, Two-dimensional Finsler metrics with constant flag curvature. Manuscripta Math. **109**, 349–366 (2002)
69. Z. Shen, Projectively flat Randers metrics with constant flag curvature. Math. Ann. **325**, 19–30 (2003)
70. Z. Shen, Projectively flat Finsler metrics of constant flag curvature. Trans. Am. Math. Soc. **355**, 1713–1728 (2003)
71. Z. Shen, Landsberg curvature, S-curvature and Riemann curvature, in *A Sampler of Finsler Geometry*, ed. by D.D.-W. Bao. MSRI Publication Series, vol. 50 (Cambridge University Press, Cambridge, 2004)
72. Z. Shen, Riemann-Finsler geometry with applications to information geometry. Chin. Ann. Math. Ser. B **27**, 73–94 (2006)
73. Z. Shen, On some non-Riemannian quantities in Finsler geometry. Can. Math. Bull. **56**, 184–193 (2013)
74. Z. Shen, G.C. Yildirim, On a class of projectively flat metrics with constant flag curvature. Can. J. Math. **60**, 443–456 (2008)
75. Z. Shen, C. Yu, On Einstein square metrics. Publ. Math. Debr. **85**, 413–424 (2014)

76. Z. Shen, G. Yang, On a class of weakly Einstein Finsler metrics. Isr. J. Math. **199**, 773–790 (2014)
77. A. Taybi, B. Najafi, On isotropic Berwald metrics. Ann. Polon. Math. **103**, 109–121 (2012)
78. D. Tang, On the non-Riemannian quantity H in Finsler geometry. Differ. Geom. Appl. **29**, 207–213 (2011)
79. Q. Xia, On a class of locally dually flat Finsler metrics of isotropic flag curvature. Publ. Math. Debr. **78**, 169–190 (2011)
80. Q. Xia, Some results on the non-Riemannian quantity H of a Finsler metric. Int. J. Math. **22**, 925–936 (2011)
81. C. Yu, On dually flat Randers metrics. Nonlinear Anal. **95**, 146–155 (2014)
82. C. Yu, H. Zhu, On a new class of Finsler metrics. Differ. Geom. Appl. **29**, 244–254 (2011)
83. Y. Yu, Y. You, Projective equivalence between an (α, β)-metric and a Randers metric. Publ. Math. Debr. **82**, 155–162 (2013)
84. L. Zhou, A local classification of a class of (α, β) metrics with constant flag curvature. Differ. Geom. Appl. **28**, 170–193 (2010)
85. L. Zhou, Projective spherically symmetric Finsler metrics with constant flag curvature in R^n. Geom. Dedicata. **158**, 353–364 (2012)
86. H. Zhu, A class of Finsler metrics of scalar curvature. Differ. Geom. Appl. **40**, 321–331 (2015)

Index

E. Guo, X. Mo, *The Geometry of Spherically Symmetric Finsler Manifolds*, SpringerBriefs in Mathematics, https://doi.org/10.1007/978-981-13-1598-5

Printed in the United States
By Bookmasters